501
Questions to Master the GED® Mathematical Reasoning Test

501 QUESTIONS TO MASTER THE GED® MATHEMATICAL REASONING TEST

LEARNINGEXPRESS®

NEW YORK

Contents

Introduction

Welcome to *501 Questions to Master the GED® Mathematical Reasoning Test!* This book is designed to provide you with tons of practice with every question type you will see on the GED Mathematical Reasoning test. It provides 501 problems so you can become familiar with the mathematical concepts you'll face on the exam, along with detailed answer explanations that go over exactly why choices are correct or incorrect.

What Topics Are Covered on the GED Mathematical Reasoning Test?

The questions on the Mathematical Reasoning test will fall into two areas: Quantitative Problem Solving and Algebraic Problem Solving.

- **Quantitative Problem Solving** math questions cover basic math concepts like multiples, factors, exponents, absolute value, ratios, percentages, averages, geometry, probability, and more. Approximately 45% of the questions will fall into this category.

- **Algebraic Problem Solving** math questions ask you to use your knowledge of the basic building blocks of math to solve problems using algebra, including linear equations, quadratic equations, functions, linear inequalities, and more. Approximately 55% of the questions will fall into this category.

What Types of Questions Are on the Mathematical Reasoning Test?

Since the GED Mathematical Reasoning test is given on a computer, you will see several different types of questions. Here are different formats of questions you will encounter on the actual test:

1. Multiple-Choice

More than 50% of the questions on the GED test will be multiple-choice. You will have to pick the best answer out of four given choices: **a**, **b**, **c**, and **d**. To select an answer, you will click your mouse in the circle next to that answer choice. To change your answer, click the circle of another answer choice. In this book, you will simply circle the correct response to a multiple-choice question.

2. Fill-in-the-Blank

For fill-in-the-blank questions, rather than being presented with a selection of possible answers from which to choose, you will need to type in an answer or answers. In this book, you can practice by writing in the correct answer on the given line or lines.

3. Drop-Down

For drop-down questions, you will need to select the correct numerical answer or phrase to complete a sentence or problem. You will click your mouse on the arrow to show all of the answer choices. Then, you will click on your chosen answer to complete the sentence, paragraph, or equation. This type of question is similar to a multiple-choice item.

4. Drag-and-Drop

To answer drag-and-drop questions, you will need to click on the correct object, hold down the mouse, and drag the object to the appropriate place in the problem, diagram, chart, or graph. In this book, you can practice this type of question by identifying which object will complete the problem, diagram, chart, or graph. Instead of dragging it, you will need to write in your answer.

5. Hot-Spot

For hot-spot questions, you will need to click on an area of the screen to indicate where the correct answer is located. For instance, you may be asked to plot a point by clicking on an empty graph. In this book, you can practice by identifying where the correct answer is located and marking the location on paper in the appropriate spot.

While it's difficult to replicate these computer items in a print book, we've dedicated a percentage of the 501 questions inside to formats other than the typical multiple-choice type.

Can I Use a Calculator?

An online calculator called TI-30XS MultiView will be available to you for most of the questions within the Mathematical Reasoning test.

The first five questions on the test will be non-calculator questions and they will make up Part 1. These questions may deal with ordering fractions and decimals, using the least common multiple (LCM) and the greatest common factor (GCF), using the distributive property, simplifying or solving problems using the rules of exponents, or identifying absolute value, among other computational skills and concepts. Part 2 of the test will have the on-screen calculator available for you to use. Most states will permit you to bring in a TI-30XS MultiView calculator, but you should check with your individual testing center regarding this. No other calculators will be permitted.

Formula Reference Sheet

A list of formulas will be available for you to use during the test. Although it will include basic formulas such as the area of a rectangle or triangle, circumference of a circle, and perimeter of geometric figures, it will benefit you greatly to be able to recall these formulas from memory and work with them comfortably without having to rely on the Formula Reference Sheet. Visit the Appendix to see the list of formulas you will be given on test day.

Area

Parallelogram: $A = bh$
Trapezoid: $A = \frac{1}{2}h(b_1 + b_2)$

Surface Area and Volume

Rectangular/right prism:	$SA = ph + 2B$	$V = Bh$
Cylinder:	$SA = 2\pi rh + 2\pi r^2$	$V = \pi r^2 h$
Pyramid:	$SA = \frac{1}{2}ps + B$	$V = \frac{1}{3}Bh$
Cone:	$SA = \pi rs + \pi r^2$	$V = \frac{1}{3}\pi r^2 h$
Sphere:	$SA = 4\pi r^2$	$V = \frac{4}{3}\pi r^3$

(p = perimeter of base B; $\pi \approx 3.14$)

Algebra

Slope of a line: $m = \frac{y_2 - y_1}{x_2 - x_1}$

Slope-intercept form of the equation of a line: $y = mx + b$

Point-slope form of the equation of a line: $y - y_1 = m(x - x_1)$

Standard form of a quadratic equation: $y = ax^2 + bx + c$

Quadratic formula: $x = \frac{-b \pm \sqrt{b^2 - 4ac}}{2a}$

Pythagorean theorem: $a^2 + b^2 = c^2$

Simple interest: $I = prt$

(I = interest, p = principal, r = rate, t = time)

How to Use This Book

This book is divided into nine chapters—each covers a specific area of math tested on the GED Mathematical Reasoning test. If you have a few months before you take your test, we suggest you work straight through the book, taking on 10 to 15 questions per night. Set aside quiet time, mute your phone, and answer the questions under the most test-like conditions possible.

If you're closer to test day and short on time, start with the specific math topics with which you feel you need the most practice. Focusing your practice on the areas where you need the most work is smart studying—the more problems you do, the more familiar your brain will get with the math format and language you will see on the exam.

The key to using this book is to make use of its answer explanations. Every question comes with a very detailed answer explanation, walking you through how to get the *right* answer, and also how you might have reached the *wrong* answer. Knowing common mistakes and how to avoid them is another key to excelling on the GED test.

The most important part of GED prep? Practice. Practice is the key to test day success, and you've made a big step in using this book to get ready for the test.

Best of luck on your GED test! Let's get started.

501
Questions to Master the GED® Mathematical Reasoning Test

1

Fractions and Decimals

The arithmetic skills surrounding fractions and decimals are the foundations of algebra. In order to be able to be successful in the more advanced sections of the GED test, it is critical that you master these building blocks.

The questions in this chapter will test you proficiency in the following areas:

- Vocabulary used to define arithmetic operations
- Least common multiples (LCMs) and greatest common factors (GCFs)
- Simplifying and creating equivalent fractions
- Operations with fractions
- Comparing and ordering decimals, fractions, and negative numbers in lists as well as on number lines

1. What is the sum of 12.03 and 4.5?
   a. 7.53
   b. 12.48
   c. 16.53
   d. 57.03

2. The table shows Ms. Kayla's GED students' results from a key-boarding quiz. What is the difference in words per minute (wpm) between the fastest- and the slowest-typing student? _____

NAME	WORDS TYPED	MINUTES
Percy	90	2
Derrick	67	2
Toneshia	84	2
Connie	70	2
Frank	59	2

3. Using the table from the previous question, how many words would Percy type in 5 minutes?
a. 450
b. 180
c. 350
d. 225

4. What expression represents 7 less than the product of 10 and a number h?
a. $7 - (10 \times h)$
b. $7 - (10 \div h)$
c. $(10 \times h) - 7$
d. $(10 \div h) - 7$

5. What is the greatest common factor of 12, 24, and 48?
a. 6
b. 8
c. 12
d. 48

6. Gina sold $\frac{5}{8}$ of a mushroom pizza and $\frac{3}{4}$ of an onion pizza to the same family. How much pizza in all did the family buy?
a. $1\frac{3}{8}$
b. $1\frac{1}{2}$
c. $\frac{8}{12}$
d. $\frac{15}{32}$

7. Katherine has $5\frac{4}{5}$ yards of red silk fabric. Steve uses $1\frac{1}{3}$ yards of it. How many yards of red silk fabric are remaining?

    a. $6\frac{5}{8}$

    b. $4\frac{3}{2}$

    c. $4\frac{7}{15}$

    d. $4\frac{2}{5}$

8. Ryan has 30 pounds of cactus sand and is making tiny pots of various succulents to sell at the Mississippi Avenue Weekend Fair. If each tiny pot requires $\frac{1}{4}$ pound of sand, how many tiny succulent pots can he prepare for the fair? _____

9. Put the following numbers in order from least to greatest: 15, 1.0005, $\frac{3}{2}$, 1.005. _____

10. Put the following list of fractions and decimals in order of least to greatest:
    $0.068, \frac{2}{3}, -1.7, \frac{8}{5}, 0.61, -1\frac{2}{3}$

(*Note:* This would be a drag and drop question on your GED test, but instead, just list them in order.)

11.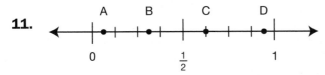

    If x is a rational number such that $\frac{1}{2} < x < \frac{3}{4}$, then which of the points on the number line above may represent x?

    a. point A
    b. point B
    c. point C
    d. point D

12. As a simplified fraction, $\frac{1}{4}(\frac{5}{2} - \frac{1}{6}) =$

   a. $-\frac{1}{4}$

   b. $\frac{1}{6}$

   c. $\frac{7}{12}$

   d. $\frac{3}{2}$

13. What expression is equivalent to the sum of $\frac{1}{2}x$ and $\frac{3}{4}x - 5$?

Select from the numbers and expressions listed here, and write the correct values in the boxes to find an equivalent expression.

   $\frac{5}{4}x$

   $\frac{2}{3}x$

   $\frac{3}{4}$

   $\frac{5}{2}$

   $\frac{1}{2}x$

14. Which of the following is equivalent to $(\frac{3}{4})^3$?

   a. $\frac{3^3}{4^3}$

   b. $\frac{3 \times 3}{4 \times 3}$

   c. $\frac{3^3}{4}$

   d. $\frac{3}{4 \times 3}$

15. Which of the following expressions is equivalent to $\frac{3}{x} \div \frac{5x}{2}$ for all nonzero x?

   a. $\frac{6}{5x^2}$

   b. $\frac{15x^2}{2}$

   c. $\frac{3}{2}$

   d. $\frac{15}{2}$

16. What is triple the difference of 65 and 50?

   a. 15

   b. 45

   c. 145

   d. 345

17. Order the following fractions on the number line: $\frac{1}{2}$, $\frac{7}{8}$, $\frac{3}{4}$.

18. What is the greatest common factor of $14x^4yz^2$ and $35xy^3$?
  a. xyz
  b. $7x^4y^3z^2$
  c. $7xy$
  d. $490x^5y^4z^2$

19. Tia has $40 less than $100. Over the week her money is increased by $80 and then doubled. How much money does she have at the end of the week?
  a. $240
  b. $280
  c. $320
  d. $440

20. Write these numbers so they are ordered from least to greatest:
    0.2, –0.045, 0.06, 0.053, –0.44

    _____

21. Richard has a T-shirt company. He needs to make at least $2,500 this month to meet his sales goal. If each T-shirt costs $16.50, what is the minimum number of T-shirts Richard needs to sell to reach his goal? _____

22. A couple traveled 2,871.46 miles cross-country and averaged 31.2 miles per gallon. How many gallons of gas did they use, rounded to the nearest tenth?
  a. 92.0
  b. 93.0
  c. 92.03
  d. 92.04

23. Consider the equation $Z = \frac{a}{\frac{1}{x} + \frac{1}{y}}$, where a, x, and y are positive real numbers. Which of the following statements is true?
   a. As x and y increase and a remains the same, Z increases.
   b. If a and x increase and y remains the same, then Z decreases.
   c. If x and y decrease and a remains the same, then Z increases.
   d. If y increases and both a and x remain the same, then Z increases.

24. The Hiking Club is organizing a school picnic for 30 people and plans on buying three hot dogs per person. If each package of hot dogs contains six hot dogs, what is the fewest number of packages the club needs to buy?
   a. 5
   b. 10
   c. 15
   d. 20

25. The pictures are drawn to show that $\frac{5}{6}$ is greater than which fraction?

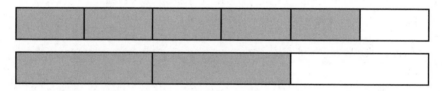

   a. $\frac{1}{6}$
   b. $\frac{1}{2}$
   c. $\frac{2}{3}$
   d. $\frac{4}{5}$

26. Which point represents $-\frac{1}{3}$ on the number line?

a. R
b. S
c. T
d. U

27. In which of the following are the numbers in order from least to greatest?
a. $0.5, 0.25, \frac{3}{5}, 2, \frac{5}{2}$
b. $0.25, 0.5, \frac{3}{5}, 2, \frac{5}{2}$
c. $0.25, 0.5, \frac{3}{5}, \frac{5}{2}, 2$
d. $0.25, \frac{5}{2}, \frac{3}{5}, 0.5, 2$

28. What is the difference between 7.90 and 0.26?
a. 6.64
b. 6.68
c. 6.74
d. 7.64

29. Cecil lives $1\frac{5}{8}$ miles from his school. Which of the following decimals is equal to the distance from Cecil's home to his school?
a. 0.625 miles
b. 1.58 miles
c. 1.125 miles
d. 1.625 miles

30. In which of the following are the numbers in order from least to greatest?
a. $6\frac{3}{5}, 6.8, 7\frac{3}{10}, 7.8, 7.9$
b. $6.8, 6\frac{3}{5}, 7\frac{3}{10}, 7.8, 7.9$
c. $7.9, 7.8, 7\frac{3}{10}, 6.8, 6\frac{3}{5}$
d. $6\frac{3}{5}, 6.8, 7.8, 7.9, 7\frac{3}{10}$

31. Mrs. Thompson buys colored sand for the class art project. There are 31 students in her class, and each student needs 0.48 pounds of sand for the project. About how many pounds of colored sand must Mrs. Thompson buy?

   a. 15 pounds

   b. 30 pounds

   c. 60 pounds

   d. 150 pounds

32. If –0.4 is an estimate of the number corresponding to one of the points labeled on the number line, which point is it?

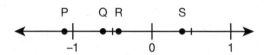

   a. P

   b. Q

   c. R

   d. S

Answers and Explanations

1. **The correct answer is choice c.** *Sum* is a key word that means add. You need to add the numbers 12.03 and 4.5. Place the numbers one over the other and line up the decimal points.

    12.03
    + 4.5

    Because 4.5 does not show as many decimal places as 12.03, add a zero on the end of 4.5 to make 4.50. Then each number will show the same number of places to the right of the decimal. These will be easier to add.

    12.03
    +4.50

    Now add each column one at a time starting on the right.

    12.03
    +4.50
    16.53

    Choice **a** is incorrect because this is the difference and not the sum of the two given values.

    Choice **b** is incorrect because you treated the 4.5 as 0.45 and lined up the decimals incorrectly.

    Choice **d** is incorrect because you treated the 4.5 as 45.0 and lined up the decimals incorrectly.

2. **The correct answer is 15.5 wpm.** To answer this question, it is helpful to recall that words per minute is a rate. First, notice that the quiz results were for 2 minutes, not 1 minute. We need to find words per minute. To do this, simply divide each number of words typed by 2 to find the words per minute for each person. Since each number is being divided by the same number, we only need to do this for the greatest number and the smallest number of words typed. $\frac{90 \text{ words}}{2 \text{ minutes}}$ = 45 wpm. $\frac{59 \text{ words}}{2 \text{ minutes}}$ = 29.5 wpm. To find the difference, simply subtract 29.5 from 45 to get 15.5 wpm. Alternatively, subtract the smallest number of words typed from the largest number and divide the difference by 2.

3. **The correct answer is choice d.** Since Percy typed 90 words in 2 minutes, divide 90 by 2 to see how many words Percy typed per minute: 90 ÷ 2 = 45 words per minute. To see how many words Percy would type in 5 minutes, multiply 45 words by 5: 225.

   Choice **a** is incorrect because you multiplied 90 by 5 minutes, but Percy types 90 words every 2 minutes and not every minute.

   Choice **b** is incorrect because Percy types 90 words every 2 minutes, so he would type 180 words in 4 minutes, not in 5 minutes.

   Choice **c** is incorrect because 350 words in 5 minutes is a rate of 70 words per minute, but Percy types at a rate of 45 words per minute, since he types 90 words in 2 minutes.

4. **The correct answer is choice c.** The phrase *product of 10 and a number h* implies that 10 and h should be multiplied in a set of parentheses. The phrase *less than* means subtraction, but you have to switch the order of the terms, so the 7 will come last: $(10 \times h) - 7$.

   Choice **a** is incorrect because the order of $(10 \times h)$ and 7 is incorrect; they must be flipped when the phrase *less than* is used.

   Choice **b** is incorrect because it uses the quotient of 10 and h and not the product. Also, the order of the two terms is incorrect as they must be flipped when the phrase *less than* is used.

   Choice **d** is incorrect because it uses the quotient of 10 and h and not the product.

5. **The correct answer is choice c.** To find the greatest common factor of 12, 24, and 48, write out the factor lists for each of these numbers:

   12: 1, 2, 3, 4, 6, <u>12</u>
   24: 1, 2, 3, 4, 6, 8, <u>12</u>, 24
   48: 1, 2, 3, 4, 6, 8, <u>12</u>, 16, 24, 48

   12 is the greatest factor in all of these lists.

   Choice **a** is incorrect because although 6 is a common factor of 12, 24, and 48, it is not the *greatest* common factor.

   Choice **b** is incorrect because 8 is not a factor of 12.

   Choice **d** is incorrect because 48 is the *least common multiple* but not the *greatest common factor* of 12, 24, and 48.

6. **The correct answer is choice a.** We need to add $\frac{5}{8}$ and $\frac{3}{4}$ in order to see how much pizza that single family bought. To find the lowest common denominator, check to see whether the smaller denominator (4) evenly divides into the larger denominator (8). Since it does, we only need to raise the smaller fraction ($\frac{3}{4}$) to higher terms to have common denominators. Multiply the numerator and denominator by 2 to raise $\frac{3}{4}$ to $\frac{6}{8}$. Then add the numerators and keep the denominator the same: $\frac{5}{8} + \frac{6}{8} = \frac{11}{8}$. This simplifies to $1\frac{3}{8}$.

   Choice **b** is incorrect because $1\frac{1}{2}$ is the sum of $\frac{3}{4}$ and $\frac{3}{4}$, but you needed to add $\frac{5}{8}$ and $\frac{3}{4}$.

   Choice **c** is incorrect because you added the given denominators instead of finding common denominators and keeping them the same. When adding fractions, keep the common denominator and only add the numerators.

   Choice **d** is incorrect this is the *product* and not the sum of the two given fractions.

7. **The correct answer is choice c.** For this question, we need to subtract the $1\frac{1}{3}$ yards of silk that Steve used from the $5\frac{4}{5}$ yards of red silk fabric that Katherine has. Use a common denominator of 15:

$$5\frac{4}{5} = 5\frac{12}{15}$$
$$1\frac{1}{3} = 1\frac{5}{15}$$

   Since the second fraction has a smaller value than the first fraction, we can perform the subtraction by subtracting the whole number and the fractions separately:

$$5\frac{12}{15} - 1\frac{5}{15} = 4\frac{7}{15}$$

   Choice **a** is incorrect because you incorrectly added the silk that Steve used to the silk that Katherine had instead of subtracting them. You added the denominators instead of finding common denominators and keeping them the same.

   Choice **b** is incorrect because you subtracted the given denominators instead of finding common denominators and keeping them the same. When subtracting fractions, find the common denominator and only subtract the numerators.

   Choice **d** is incorrect because $\frac{4}{5} - \frac{1}{3}$ is $\frac{7}{15}$ and not $\frac{2}{5}$.

8. **The correct answer is 120.** Our task is to take 30 pounds of soil and divide it into equal portions of $\frac{1}{4}$ pound each. To do this, we need to perform $30 \div \frac{1}{4}$. When dividing with fractions, multiply the first number by the reciprocal of the second fraction: $30 \div \frac{1}{4} = 30 \times \frac{4}{1}$ = 120. Ryan will be able to make 120 tiny succulent pots for the fair.

9. **The correct answer is 1.0005, 1.005, $\frac{3}{2}$, 15.** All of these numbers, when rewritten, contain the digits 1 and 5. The number 1.005 = 1$\frac{5}{1,000}$, the number 1.0005 = 1$\frac{5}{10,000}$, and $\frac{3}{2} = 1\frac{1}{2} = 1.5$. Therefore, 1.0005 is less than 1.005, which, in turn, is less than 1.5. The greatest number is 15.

10. **The correct answer is –1.7, –1$\frac{2}{3}$, 0.068, 0.61, $\frac{2}{3}$, $\frac{8}{5}$.** Convert the fractions into decimals and add zeros so that they all have three places to the right of the decimal. Written in this fashion, they are all in *thousandths* and can be compared:

    $0.068 = \frac{68}{1,000}$

    $\frac{2}{3} = 0.667 = \frac{667}{1,000}$

    $-1.7 = -1.700 = \frac{-1,700}{1,000}$

    $\frac{8}{5} = 1.600 = \frac{1,600}{1,000}$

    $0.61 = 0.610 = \frac{610}{1,000}$

    $-1\frac{2}{3} = -1.667 = \frac{-1,667}{1,000}$

    Since all the fractions above are now out of 1,000, we can ignore the denominators and just focus on the numerators: –1,700, –1,667, 68, 610, 667, 1,600.

11. **The correct answer is choice c.** The middle hash mark between one-half and one represents three-fourths. Point C is between this mark and the one-half mark, indicating it satisfies the given inequality.

    Choice **a** is incorrect. This point is much smaller than one-half. In fact, it is smaller than one-fourth.

    Choice **b** is incorrect. This point is between one-fourth and one-half.

    Choice **d** is incorrect. This point is larger than three-fourths.

12. **The correct answer is choice c.** $\frac{1}{4}(\frac{5}{2} - \frac{1}{6}) = \frac{1}{4}(\frac{15}{6} - \frac{1}{6}) = \frac{1}{4}(\frac{14}{6}) = \frac{7}{12}$.
Choice **a** is incorrect. Denominators are never subtracted when subtracting two fractions.
Choice **b** is incorrect. When rewriting the first fraction with the common denominator of 6, the numerator must also be multiplied by 2.
Choice **d** is incorrect. Parentheses indicate multiplication, not addition. Further, addition of fractions doesn't involve adding the denominators.

13. **The correct answer is $\frac{5}{4}x - 5$.** In order to find the sum of the two given expressions, add the like terms $\frac{1}{2}x$ and $\frac{3}{4}x$ by finding common denominators, and keeping the constant, –5, on its own: $\frac{1}{2}x + \frac{3}{4}x - 5 = \frac{2}{4}x + \frac{3}{4}x - 5 = \frac{5}{4}x - 5$.

14. **The correct answer is choice a.** Applying an exponent to a fraction is equivalent to applying that exponent to the numerator and the denominator.
Choice **b** is incorrect. An exponent of 3 is not equivalent to multiplication by 3.
Choice **c** is incorrect. The exponent must be applied to both the numerator and the denominator.
Choice **d** is incorrect. An exponent of 3 is not equivalent to multiplication by 3 and would be applied to both the numerator and the denominator.

15. **The correct answer is choice a.** The division is equivalent to $\frac{3}{x} \times \frac{2}{5x} = \frac{6}{5x^2}$.
Choice **b** is incorrect. The division of two fractions is equivalent to multiplying the first fraction by the reciprocal of the second fraction.
Choice **c** is incorrect. This is the result of multiplying and not dividing the fractions if the 5 canceled out. There are no terms that would cancel with the 5.
Choice **d** is incorrect. This is the result of multiplying the two fractions.

16. The correct answer is choice b. The word *triple* indicates that the following expression should be grouped within parentheses. *Difference* means to subtract. Therefore, translate *triple the difference of 65 and 50* into $3 \times (65 - 50)$, which simplifies to $3 \times 15 = 45$.

Choice **a** is incorrect because 15 is only the difference of 65 and 50, but you forgot to triple it.

Choice **c** is incorrect because you multiplied 3 by 65 first, and then subtracted 50, but you needed to find the difference first before multiplying.

Choice **d** is incorrect because 345 is triple the *sum* of 65 and 50. You needed to subtract 50 from 65, rather than add them.

17.

First, rewrite the fractions so that the denominator (the bottom number) of each fraction is the same. To do this, look for what number all three numbers—2, 8, and 4—can go into. The answer is 8. One of the fractions already has a denominator of 8, so that fraction is good to go. Changing the denominator of the other two fractions, $\frac{1}{2}$ becomes $\frac{4}{8}$, and $\frac{3}{4}$ becomes $\frac{6}{8}$. Now it is easy to see that $\frac{4}{8}$ is smaller than $\frac{6}{8}$, which is smaller than $\frac{7}{8}$.

18. The correct answer is choice c. First, determine that the greatest common factor of 14 and 35 is 7 by making factors lists for each number and selecting the greatest factor in common: 14: 1, 2, 7, 14 and 35: 1, 5, 7, 35. Next, determine the greatest power of each variable that exists in *both* expressions. There is only x^1 and y^1 in common in both expressions, so the greatest common factor is $7xy$.

Choice **a** is incorrect because 7 is the greatest common factor of 14 and 35, and there is not a z in $35xy^3$, so z cannot be in the greatest common factor.

Choice **b** is incorrect because you chose the greatest power of each variable in the two expressions, not the greatest power they had in common.

Choice **d** is incorrect because it is the product of $14x^4yz^2$ and $35xy^3$, not the greatest common factor.

19. **The correct answer is choice b.** First, interpret *$40 less than $100* to mean $100 – $40, since *less than* means subtraction in the opposite order. Increasing that by $80 means adding $80 to $100 – $40. Finally, *and then doubled* is asking the $100 – $40 + $80 to be put in parentheses before multiplying it by 2, so that everything is doubled, and not just the first or last term: 2(100 – 40 + 80) = 280.
Choice **a** is incorrect because you doubled $100 and added $40, instead of subtracting $40 from $100 and adding $80.
Choice **c** is incorrect because you doubled the sums of $100 and $80, and *then* subtracted $40, when you needed to subtract $40 first.
Choice **d** is incorrect because you added the original $40 to $100, instead of subtracting it, before adding $80 and doubling it.

20. **The correct answer is –0.44, –0.045, 0.053, 0.06, 0.2.** In order to arrange these from least to greatest, first add zeros to the ends of the decimals so that each number is expressed as thousandths.

    0.2 = 0.200
    –0.045 = –0.045
    0.06 = 0.060
    0.053 = 0.053
    –0.44 = –0.440

Now since these are all thousandths, the decimals can be ignored and you can compare these numbers as 200, –45, 60, 53, and –440, which would order as –440, –45, 53, 60, 200, so the correct ordering is: –0.44, –0.045, 0.053, 0.06, 0.2.

21. **The correct answer is 152.** To translate these words into symbols
so we can solve the problem, we first need to identify the unknown.
What is it that we are trying to find out? The number of T-shirts.
This is what then becomes represented by a variable. So, let's say t is
the number of T-shirts.

We know that Richard needs to make at least $2,500. What do the
words *at least* sound like? Could he make more? We're sure he'd
love to! So, the words *at least* indicate that we have an inequality on
our hands. The total amount Richard makes should be equal to *or*
greater than $2,500.

How do we calculate how much money Richard will make? He gets
$16.50 per T-shirt sold. So, we are going to be multiplying $16.50
by our variable t, which represents the number of T-shirts, to get a
product of $16.50t$.

Putting everything together, we now get the mathematical statement
$2,500 \le 16.50t$. The inequality sign is pointing to 2,500 because
Richard wants more than $2,500 if he can sell that many shirts!

We solve this inequality just like we would if there were an equal
sign. Divide both sides by 16.50 to undo the multiplication of $16.50t$.

$$\frac{2,500}{16.50} \le \frac{16.50t}{16.50}$$

Now, $\frac{16.50}{16.50} = 1$, so we now have t by itself on one side of the inequality. On the other side of the inequality, we can simplify and get
151.5. This number represents the number of T-shirts, so we need
to round up to get a whole number, 152. Therefore, Richard needs
to sell at least 152 T-shirts to make his goal of $2,500. Mathemati-
cally, this is written $152 \le t$, as we hope the number of T-shirts, t, is
greater than 152. However, t needs to be *at least* equal to 152 for
Richard to make $2,500.

22. The correct answer is choice a. Divide 2,871.46 by 31.2 to get 92.03397436. In order to round to the tenths place, use the 3 in the hundredths place as an indicator that the tenths place will remain the same. This yields 92.0.

Choice **b** is incorrect because you rounded to the ones digit incorrectly. The digit to its immediate right would need to be 5 or greater.

Choice **c** is incorrect because you rounded to the hundredths place and not to the tenths place.

Choice **d** is incorrect because you rounded to the hundredths place and did so incorrectly.

23. The correct answer is choice d. As y increases, its reciprocal $\frac{1}{y}$ decreases. Therefore, you would be dividing by less in the expression for Z, which makes its value increase.

Choices **a** and **b** are incorrect because you are dividing by less, which means the value of Z would increase.

Choice **c** is incorrect because as x and y decrease, their reciprocals increase. So you are dividing by more, which means the value of Z would decrease.

24. The correct answer is choice c. Each person is expected to eat three hot dogs, so the total number of hot dogs needed for 30 people is 30 × 3 = 90 hot dogs. If each package contains six hot dogs, the fewest number of packages the Hiking Club can buy is 90 ÷ 6 = 15 packages.

Choice **a** is incorrect because five packages of six hot dogs in a package would equal a total of 30 hot dogs, which is enough for only one hot dog per person.

Choice **b** is incorrect because ten packages of six hot dogs in a package would equal a total of 60 hot dogs, which is enough for only two hot dogs per person.

Choice **d** is incorrect because it is not the fewest number of packages needed for each person to have three hot dogs. Twenty packages of six hot dogs in a package would equal a total of 120 hot dogs, which means that each person could have four hot dogs.

25. The correct answer is choice c. Notice that both strips are the same length. Since the first strip is divided into six sections, it is divided into sixths. Since 5 out of the 6 sections are shaded, it shows $\frac{5}{6}$. Since the second strip is divided into three sections, it is divided into thirds. Since 2 out of the 3 sections are shaded, it shows $\frac{2}{3}$. The shaded part of the top strip is longer than the shaded part of the lower strip, showing that $\frac{5}{6}$ is longer, or greater, than $\frac{2}{3}$.

26. The correct answer is choice c. Point T is $\frac{1}{3}$ of the way between 0 and −1.

Choice **a** is incorrect because R is at $-1\frac{1}{3}$.

Choice **b** is incorrect because S is at $-\frac{2}{3}$.

Choice **d** is incorrect because U is at $\frac{1}{3}$.

27. The correct answer is choice b. The numbers 0.25, 0.5, and $\frac{3}{5}$ are less than 1, and 2 and $\frac{5}{2}$ are greater than 1. The positions of these numbers are shown on the number line.

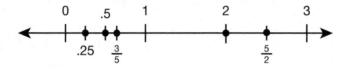

Choice **a** is incorrect because 0.5 (or 0.50) is greater than 0.25.

Choice **c** is incorrect because $\frac{5}{2}$ = 2.5, and 2.5 > 2.

Choice **d** is incorrect because $\frac{5}{2}$ is greater than 2, and also greater than $\frac{3}{5}$, so $\frac{5}{2}$ should be to the right of both numbers.

28. **The correct answer is choice d.** The key word *difference* means subtract. Begin by writing the numbers one over the other, lining up the decimal points.

$$7.90$$
$$\underline{-0.26}$$

Because 6 is larger than 0 in the hundredths place, borrow 10 from the 9 in the tenths place. This makes the 9 an 8 and the 0 a 10.

$$7.\overset{8\ 10}{\cancel{9}\cancel{0}}$$
$$\underline{-0.26}$$

Now subtract the columns.

$$7.\overset{8\ 10}{\cancel{9}\cancel{0}}$$
$$\underline{-0.26}$$
$$7.90$$

The answer is 7.64.

29. **The correct answer is choice d.** Convert the fraction to a decimal by dividing 5 by 8:

$$5 \div 8 = 0.625$$

Cecil lives $1\frac{5}{8}$ miles from his school, so place a 1 in the ones place of the decimal—1.625.

30. **The correct answer is choice a.** The fractions can easily be converted to decimals. The first decimal place to the right of the decimal point is the tenths place. So $\frac{3}{10}$ (or three-tenths) = 0.3, and $7\frac{3}{10}$ = 7.3. The fraction $\frac{3}{5}$ can be converted into tenths: $\frac{3}{5} \times \frac{2}{2} = \frac{6}{10}$. Thus, $6\frac{3}{5}$ = 6.6. Now that all of the numbers are in the same form, they can be easily ordered from least to greatest.

Choice **b** is incorrect because 6.8 is not less than $6\frac{3}{5}$.

Choice **c** is incorrect because the numbers are ordered from greatest to least.

Choice **d** is incorrect because $7\frac{3}{10}$ is not the greatest number.

31. The correct answer is choice a. Round the numbers in the problem to help you estimate the answer.

Rounded to the nearest 10, there are 30 students in Mrs. Thompson's class. Rounded to the nearest tenth of a pound, each student
needs about 0.5, or $\frac{1}{2}$ pound of sand.

Multiply the number of students by the amount of sand each student
needs:

30 students × 0.5 pounds of sand per student = 15 pounds

Choice **b** is incorrect because 30 pounds is too high—0.48 is less
than half a pound of sand per student. Rounding 0.48 to 1.00
would not be a good estimate.

Choice **c** is incorrect because you must multiply 30 by 0.5, not
divide.

Choice **d** is incorrect because 150 pounds is much too high. Be careful with the decimal point—you must multiply 30 by 0.5, not 5.

32. The correct answer is choice c. Look at the labels on the number
line from left to right. There is a single unlabeled tick mark halfway
between –1 and 0 and another halfway between 0 and 1. This means
the tick marks represent –0.5 and 0.5, respectively.

Since –0.4 is between 0 and –0.5, the point should be between 0 and
the first tick mark to the left of 0. The only point that meets this cri-
terion is R.

Choices **a** and **b** are incorrect because points P and Q are both less
than -0.5, rather than between –0.5 and 0.

Choice **d** is incorrect—since the point is supposed to be negative,
immediately eliminate this choice.

2

Negatives, Exponents, and Order of Operations

Stock prices fall, money gets spent, and temperatures drop—all of these are illustrations of negative numbers in action, so one should be adept at working with non-positive numbers. Similarly, financial investments increase at exponential rates and the amount of carbon in fossils decreases at exponential rates, so understanding exponents is another important skill. In order to be successful on the GED test, you must be comfortable working the negative numbers and exponents, and you must know how to correctly execute the order of operations when evaluating expressions. The questions in this chapter will test your proficiency in the following areas:

- Rules for working with signed numbers
- Order of operations (PEMDAS)
- Laws of exponents
- Square roots and cube roots
- Absolute value
- Scientific notation
- Recognizing rational, irrational, and undefined numbers

33. Evaluate the following expression:

$(5 - 3) \times (4 + 4 \div 2)$

a. 6

b. 8

c. 10

d. 12

34. Simplify the following problem: $\sqrt{\frac{75}{72}}$.

a. $\dfrac{5\sqrt{3}}{6\sqrt{2}}$

b. $\dfrac{3\sqrt{5}}{8\sqrt{3}}$

c. $\dfrac{25 \cdot 3}{8 \cdot 9}$

d. $\sqrt{\dfrac{75}{72}}$

35. Simplify. $\sqrt[3]{-1,000} - 3^2$ _____

36. Which absolute value expression illustrates the distance between point A and point B on the following number line?

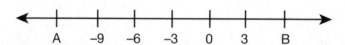

a. $|12 - 6|$

b. $|-12 + 6|$

c. $|-12 - 6|$

d. $|6 - 12|$

37. What is the value of h on the following number line?

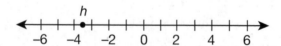

a. $-4\frac{1}{2}$

b. $-3\frac{1}{2}$

c. $4\frac{1}{2}$

d. $3\frac{1}{2}$

38. Which of the following is the number 316.72 written in scientific notation?

    a. 3.1672×10^{-2}

    b. 3.1672×10^{2}

    c. 3.1672×10^{3}

    d. 3.1672×10^{1}

39. Pluto is 5,914,000,000 km from the sun. This distance can be written in scientific notation as

    a. 59.14×10^{8}

    b. 5.914×10^{9}

    c. 0.5914×10^{10}

    d. 5.914×10^{6}

40. Which of the following is an irrational number?

    a. $\sqrt{12}$

    b. $\sqrt{9}$

    c. $\frac{7}{3}$

    d. $\frac{0}{11}$

41. For what two values of x is the following numerical expression undefined? $\frac{12x}{x^2 - 25}$. _____

42. Which of the following is equivalent to $\frac{2^5}{2^2}$?

    a. 2

    b. 2^{3}

    c. 2^{7}

    d. 2^{10}

43. What is the value of the expression $-3x + 10y$ when $x = -4$ and $y = -2$?

    a. -34

    b. -32

    c. -8

    d. 1

44. What is the value of $\frac{x - 5}{x^2 + 1}$ when $x = -3$?

   a. $-\frac{3}{2}$

   b. $-\frac{4}{5}$

   c. $\frac{8}{5}$

   d. 1

45. Which of the following is equivalent to $\frac{\sqrt[3]{9} \times \sqrt[3]{18}}{3}$?

   a. $\sqrt[3]{2}$

   b. $3\sqrt[3]{2}$

   c. $\sqrt[3]{6}$

   d. $\sqrt[3]{18}$

46. Write your answer in the box. You may use numbers, symbols, and/or text in your response.

Simplify the following expression completely. Be sure to leave your answer in radical form.

$$\frac{\sqrt{72}}{\sqrt{36}}$$

┌─────────────────────────────────────┐
│   │
└─────────────────────────────────────┘

47. Which of the following is equivalent to $5^{\frac{1}{2}} \times 5^2$?

   a. $5^{-\frac{3}{2}}$

   b. 5

   c. $5^{\frac{5}{2}}$

   d. $5^{\frac{1}{4}}$

48. A specialized part for a manufacturing process has a thickness of 1.2×10^{-3} inches. To the ten-thousandth of an inch, what would be the thickness of a stack of ten of these parts?

   a. 0.0001

   b. 0.0012

   c. 0.0120

   d. 0.1200

49. Which of the following is equivalent to the numerical expression $\sqrt{2}(\sqrt{18} - \sqrt{6})$?

    a. $4\sqrt{3}$
    b. $5\sqrt{6}$
    c. $6 - 2\sqrt{3}$
    d. $6 - \sqrt{6}$

50. The product of $x^2 - 6$ and x^4 is

    a. $x^8 - 6$
    b. $x^6 - 6$
    c. $x^6 - 6x^4$
    d. $x^8 - 6x^4$

51. What is the value of $\frac{x-5}{x^2-1}$ when $x = \frac{1}{2}$?

    a. -10
    b. $\frac{3}{2}$
    c. 6
    d. 0

52. A new computer can perform 2.9×10^9 calculations per second, whereas last year's model could perform only 3.2×10^8 calculations per second. How many more calculations can this year's model perform compared to last year's model?

    a. 0.3×10^1
    b. 2.58×10^9
    c. 0.3×10^9
    d. 3.22×10^9

53. Fill in the box with the appropriate exponent: $\dfrac{\sqrt{x}}{x\sqrt[3]{x}} = x^{\square}$

    a. $\frac{11}{6}$
    b. $-\frac{5}{6}$
    c. -2
    d. $\frac{3}{8}$

54. Which of the following is equivalent to $\frac{2^3 \times 7^3}{14^{-2}}$?
   a. 14^{-5}
   b. 14^1
   c. 14^8
   d. 14^5

55. What is 0.00231 written in scientific notation?
   a. 231×10^{-3}
   b. 231×10^{-5}
   c. 2.31×10^{-3}
   d. 2.31×10^3

56. Which of the following expressions represents the expression $64 - 8x$ in lowest terms using the distributive property?
   a. $4(16 - 2x)$
   b. $8 \cdot 8 + 8 \cdot x$
   c. $8^2 + 8^1 x$
   d. $(8 - x)^8$

57. Which of the following expressions represents $\sqrt{20} \cdot \sqrt{12}$ in lowest terms?
   a. $10 \cdot 6$
   b. $\sqrt{240}$
   c. $4\sqrt{15}$
   d. $16\sqrt{15}$

58. What is $2x^2 + x - 4$ when $x = -3$?
   a. 11
   b. 17
   c. 29
   d. -25

59. Evaluate $-3 + -10(-5 - 14)$. _____

60. Find the value of the following expression: $(-5)^2 - (-4)^3$.
   a. 2
   b. -39
   c. 39
   d. 89

61. Simplify: $\frac{4 + 3 \times 2}{7 - 14 \div 7}$.
   a. 2
   b. $\frac{14}{5}$
   c. 14
   d. −14

62. The sun is 93,000,000 miles from Earth. Which of the following expressions represents this distance in scientific notation?
   a. 93×10^6
   b. 93×10^5
   c. 9.3×10^7
   d. 0.93×10^8

63. Simplify: $|-20 + 5| - |-40|$ _____

64. Evaluate the following expression for the value of $x = -2$:
$4x^2 + 3(1 - x)$.
   a. 21
   b. 25
   c. 73
   d. 79

65. What is the sum of the greatest and least of the following values?
$\frac{2^{-2}}{3}$, $(-3)^2$, $(16)^{-1}(\sqrt{5})^4$, $(-2)^3$
   a. 17
   b. 27.25
   c. 1
   d. $13\sqrt{5}$

66. Evaluate the expression $\frac{(t + 6) - 5}{t^2 + t - 7}$ when $t = -3$.
   a. 2
   b. −8
   c. $\frac{4}{5}$
   d. −2

67. Evaluate the following expression:

$2 \times (3 \times 4 + 6 \div 3)$

   a. 10

   b. 12

   c. 20

   d. 28

68. What is the value of $(-2)^5$?

   a. -32

   b. -25

   c. 25

   d. 32

69. Evaluate the following expression, and express the answer in scientific notation:

$(5 \times 10^3) \times (3 \times 10^4)$

   a. 8×10^7

   b. 8×10^{12}

   c. 1.5×10^8

   d. 15×10^{12}

70. Which expression is equivalent to $\sqrt{108} \cdot \sqrt{32}$?

   a. $6\sqrt{3}$

   b. $4\sqrt{2}$

   c. $24\sqrt{6}$

   d. $24\sqrt{5}$

71. Which expression illustrates all three relationships specified here?

   1. $5 \cdot 5 \cdot 5 \cdot 5 \cdot 5 - 2$

   2. $\sqrt{3} \cdot \sqrt{3} \cdot \sqrt{3} \cdot \sqrt{3} \cdot \sqrt{3} - 2$

   3. $(46 - 2)(46 - 2)(46 - 2)(46 - 2)(46 - 2) - (98 - 96)$

   a. $n^4(n - 2)$

   b. $n^5 - 2$

   c. n^5

   d. $2 - n^5$

72. Which of the following is a rational number?

    a. $\frac{\pi}{\sqrt{2}}$

    b. $\frac{\frac{1}{6}}{\left(\frac{2}{3}\right)^{-4}}$

    c. $\sqrt[3]{2} \cdot \sqrt{2}$

    d. $(3\pi)^2$

73. Fill in the box with the correct exponent:

$$\frac{\sqrt{x^3}}{\sqrt[4]{x}} = x^{\square}$$

    a. 6

    b. $\frac{14}{3}$

    c. $-\frac{10}{3}$

    d. $\frac{5}{4}$

74. Which of the following expressions is equivalent to
$3x(1 - 2x) + 4x^2(x - 3) - 2x^3$?

    a. $2x^3 + x - 3$

    b. $2x^3 - 18x^2 + 3x$

    c. $2x - 6x^2 + 2x^3$

    d. $2x + 18x^2 + 2x^3$

75. The half-life of a radioactive isotope X is 2.3×10^4 years, and the half-life of a different radioactive isotope Y is 3.1×10^5 years. How many years greater is the half-life of Y than the half-life of X?

    a. 3.33×10^9 years

    b. 0.8×10^1 years

    c. 3.33×10^5 years

    d. 2.87×10^5 years

76. Which of the following expressions is equivalent to $2^4 \times 25^2$?

    a. $(2 \times 25)^{4 \times 2}$

    b. $(2 \times 5)^4$

    c. $(2 \times 25)^{2 + 4}$

    d. $(2 \times 5)^{4 + 4}$

77. Five experimenters derive different expressions for the concentration, C, of a certain toxic chemical in well water in a neighborhood over time t. Which of the following expressions is NOT equivalent to all of the others?

   a. $C = 57(\frac{2}{3})^{t+1}$

   b. $C = \frac{171}{2}(\frac{2}{3})^{t}$

   c. $C = 38(\frac{2}{3})^{t}$

   d. $C = 38(\frac{3}{2})^{-t}$

78. Fill in the box with the correct exponent: $\sqrt[3]{x^2} \cdot \sqrt{x^5} = x^{\square}$

   a. $\frac{5}{3}$

   b. $\frac{19}{6}$

   c. $\frac{19}{10}$

   d. $\frac{3}{5}$

79. Light travels at an approximate rate of 3.2×10^8 meters per second. How far does light travel in one day?

   a. 7.68×10^9 meters

   b. 1.92×10^{10} meters

   c. 1.935×10^{14} meters

   d. 2.7648×10^{13} meters

80. Fill in the box with the correct exponent: $\sqrt[3]{\sqrt{\frac{1}{5}}} = 5^{\square}$

   a. $\frac{1}{6}$

   b. $-\frac{5}{6}$

   c. -6

   d. $-\frac{1}{6}$

Answers and Explanations

33. The correct answer is choice d. Order of operations says that you should first perform any operations in parentheses, and then perform all multiplication and division, moving from left to right; then perform all addition and subtraction, again moving from left to right. Start by simplifying the expressions inside both sets of parentheses, and in the second expression, be sure to divide 4 by 2 before adding 4:

$$(5 - 3) \times (4 + 4 \div 2) = 2 \times (4 + 2)$$
$$= 2 \times 6$$
$$= 12$$

Choice **a** is incorrect because 6 is the answer to the second set of parentheses, $(4 + 4 \div 2)$, but you forgot to multiply that by the difference of 2 found in the first set of parentheses, $(5 - 3)$.

Choice **b** is incorrect. You correctly determined that $5 - 3$ is 2 and that $(4 + 4 \div 2) = 6$, but then you added these two quantities together instead of multiplying them.

Choice **c** is incorrect because in the second parentheses, $(4 + 4 \div 2)$, you incorrectly performed $4 + 4$ before doing $4 \div 2$. The order of operations says that you must perform division before addition.

34. The correct answer is choice a. To simplify this expression, knowledge of the laws of roots is needed. The square root of a fraction is equivalent to the square root of the numerator and the square root of the denominator: $\frac{\sqrt{75}}{\sqrt{72}}$. It is also important to recognize that if we rewrite each term as a product of two factors, we may be able to further simplify. $\frac{\sqrt{75}}{\sqrt{72}}$ can be written as $\frac{\sqrt{25} \cdot \sqrt{3}}{\sqrt{9} \cdot \sqrt{8}}$. This can be further simplified because 25 and 9 are both perfect squares. $\frac{\sqrt{25} \cdot \sqrt{3}}{\sqrt{9} \cdot \sqrt{8}} = \frac{5\sqrt{3}}{3\sqrt{8}}$. $\frac{5\sqrt{3}}{3\sqrt{8}}$ can be further simplified by pulling the perfect square, 4, out of $\sqrt{8}$ in the denominator: $\frac{5\sqrt{3}}{3\sqrt{8}} = \frac{5\sqrt{3}}{3\sqrt{4}\sqrt{2}} = \frac{5\sqrt{3}}{3 \cdot 2\sqrt{2}} = \frac{5\sqrt{3}}{6\sqrt{2}}$. (Note that you could have used the perfect square 36 in your first simplification: $\sqrt{\frac{75}{72}} = \frac{\sqrt{25}\sqrt{3}}{\sqrt{36}\sqrt{2}} = \frac{5\sqrt{3}}{6\sqrt{2}}$.)

Choice **b** has the square root signs assigned to the wrong numbers.
Choice **c** reflects a factorization of 75 and 72 but lost the square root.
Choice **d** suggests that this problem cannot be simplified, when in fact it can.

35. **The correct answer is –19.** The cubed root of –1,000 is –10, since $-10 \times -10 \times -10 = -1,000$. 3^2 is equal to 9, so rewrite $\sqrt[3]{-1,000} - 3^2$ as $-10 - 9$. –19 is the final answer.

36. **The correct answer is choice c.** Remember that the absolute value of the number is the distance between that number and 0. The scale of the number line is 3-unit increments. Thus, point A is –12 and point B is 6. To find the difference between these two numbers, simply make a subtraction problem and take the absolute value.

Choice **a** is incorrect because it models the distance between 12 and 6, not between –12 and 6.

Choice **b** is incorrect because –12 + 6 is the same as –12 – (-6), which models the distance between –12 and –6, not between –12 and 6.

Choice **d** is incorrect because it models the distance between 6 and 12, not between –12 and 6.

37. **The correct answer is choice b.** Even though only even numbers are drawn into this number line, we can see that each tick mark represents 1. h is halfway between the tick marks for –3 and –4, so the value for h must be $-3\frac{1}{2}$. Answer choices **c** and **d** are incorrect since these positive numbers sit on the right side of the 0 on the number line. Answer choice **a** is not correct because if h had a value of $-4\frac{1}{2}$, it would have to be sitting on the left side of 4 on the number line.

38. **The correct answer is choice b.** Scientific notation expresses a number as the product of a number between 1 and 10, including 1 but excluding 10, and a power of 10. If the number is greater than 1, then the exponent of 10 is non-negative. So, to write 316.72 in scientific notation, move the decimal point two places to the left to get a number between 1 and 10, and write the power of 10 as 2 because you moved the decimal point two places to the left.

Choice **a** is incorrect since this is the scientific notation for 0.031672 since the exponent of –2 moves the decimal point to the left twice.

Choice **c** is incorrect since this is the scientific notation for 3,167.2 since the exponent of 3 moves the decimal point to the right three times.

Choice **d** is incorrect since this is the scientific notation for 31.672 since the exponent of 1 moves the decimal point to the right once.

39. The correct answer is choice b. A number in scientific notation is written as a number that is at least 1 but less than 10, multiplied by a power of 10. The power of 10 is the number of places that the decimal is moved to transform the number into decimal notation (regular numbers). If the decimal point in the number 5.914 is moved nine places to the right, the number becomes 5,914,000,000.

Choices **a** and **c** are incorrect because although in each case the product given represents the same numerical value as the given answer, the first number is not between 1 and 10, so it is not written in scientific notation.

Choice **d** is incorrect because you confused the exponent of 6 as an indication to remove the 6 zeros, when what you needed to do was to move the decimal point nine times to the right, until the number is between 1 and 10.

40. The correct answer is choice a. Irrational numbers cannot be written as a fraction, so we know that answer choices **c** and **d** are incorrect. (Remember that $\frac{0}{11}$ has a value of 0.) Irrational numbers have a decimal value that does not terminate or repeat. Since the value of $\sqrt{9}$ is 3, this means that choice **b** can also not represent an irrational number. Therefore, $\sqrt{12}$ is the only possible choice for an irrational number.

41. The correct answer is −5 and 5. We are concerned only with the denominator when talking about undefined expressions. We need to find the two values of x that make the denominator equal to zero. Let's set up an equation and solve for x:

$$x^2 - 25 = 0$$
$$x^2 = 25$$
$$\sqrt{x^2} = \sqrt{25}$$
$$x = 5 \text{ and } -5$$

Remember, a square root is the number that, when multiplied by itself, gives you the number you started with. In this problem, we are looking for the number that when multiplied by itself yields 25, which is 5. Also, when you square a negative number, you get a positive answer. So, −5 × −5 *also* equals 25. Therefore, the two values of x that make the equation undefined are 5 and −5.

42. The correct answer is choice b. By the laws of exponents,

$$\frac{2^5}{2^2} = 2^{5-2} = 2^3.$$

Choice **a** is incorrect. When subtracted according to the laws of exponents, there will be a final exponent larger than 1.

Choice **c** is incorrect. The laws of exponents require subtraction here instead of addition.

Choice **d** is incorrect. The laws of exponents require subtraction here instead of multiplication.

43. The correct answer is choice c. $-3(-4) + 10(-2) = 12 - 20 = -8$.

Choice **a** is incorrect. This results from mixing up the substitutions of x and y. The term multiplied by -3 should be -4.

Choice **b** is incorrect. The product of -3 and -4 is positive since both signs are negative.

Choice **d** is incorrect. When substituting values into the expression, the notations $-3x$ and $10y$ indicate multiplication and not addition.

44. The correct answer is choice b. $\frac{(-3) - 5}{(-3)^2 + 1} = \frac{-8}{10} = -\frac{4}{5}$.

Choice **a** is incorrect. The numerator of the fraction shows subtraction of 5 from x, not multiplication.

Choice **c** is incorrect. The value of $(-3)^2$ is 9, not 6.

Choice **d** is incorrect. The value of $(-3)^2$ is 9, not -9.

45. The correct answer is choice c. The product in the numerator can be written as $\sqrt[3]{3 \times 3 \times 3 \times 6} = 3\sqrt[3]{6}$. The 3 in the denominator cancels out the 3 in front of the root.

Choice **a** is incorrect. The numerator is made up of a product. The denominator can cancel only one factor of the numerator.

Choice **b** is incorrect. The denominator cannot cancel out a factor within a cube root.

Choice **d** is incorrect. The cube root of 9 is not 3.

46. The correct answer is $\sqrt{2}$. Two factors of 72 are 2 and 36.

Further, $\frac{\sqrt{a}}{\sqrt{b}} = \sqrt{\frac{a}{b}}$ for positive numbers a and b.

Using these properties, $\frac{\sqrt{72}}{\sqrt{36}} = \frac{\sqrt{2 \times 36}}{36} = \sqrt{2}$.

47. **The correct answer is choice c.** When multiplying terms with the same base, the exponents are added.

Therefore $5^{\frac{1}{2}} \times 5^2 = 5^{\frac{1}{2}+2} = 5^{\frac{1}{2}+\frac{4}{2}} = 5^{\frac{5}{2}}$.

Choice **a** is incorrect. When multiplying terms with the same base, the exponents are added, not subtracted.

Choice **b** is incorrect. When multiplying terms with the same base, the exponents are added, not multiplied.

Choice **d** is incorrect. When multiplying terms with the same base, the exponents are added, not divided.

48. **The correct answer is choice c.** $1.2 \times 10^{-3} = 0.0012$ and $10 \times 0.0012 = 0.0120$.

Choice **a** is incorrect. It is not possible for the thickness of ten parts to be smaller than the thickness of one part.

Choice **b** is incorrect. This is the thickness of a single part.

Choice **d** is incorrect. This is the thickness of a stack of 100 such parts.

49. **The correct answer is choice c.** Distributing the square root of 2 and simplifying:

$$\sqrt{2}(\sqrt{18} - \sqrt{6}) = \sqrt{36} - \sqrt{12} = 6 - \sqrt{4 \times 3} = 6 - 2\sqrt{3}$$

Choice **a** is incorrect. Radicals and whole numbers are not like terms and therefore cannot be combined.

Choice **b** is incorrect. The square root of 2 must be distributed to both terms. Additionally, the radical and the remaining whole number are not like terms.

Choice **d** is incorrect. The square root of 2 must be distributed to both terms in the parentheses.

50. **The correct answer is choice c.** The two steps are to distribute and add exponents.

$$x^4(x^2 - 6) = x^{4+2} - 6x^4 = x^6 - 6x^4.$$

Choice **a** is incorrect. When two terms with the same base are multiplied, their exponents are added. Further, the term x^4 must be distributed to every term in the given binomial $x^2 - 6$.

Choice **b** is incorrect. The term x^4 must be distributed to every term in the given binomial $x^2 - 6$.

Choice **d** is incorrect. When two terms with the same base are multiplied, their exponents are added.

51. The correct answer is choice c. After plugging in the given value of x, we must simplify the result using basic operations with fractions:

$$\frac{\frac{1}{2} - 5}{\frac{1}{4} - 1} = \frac{\frac{1}{2} - \frac{10}{2}}{\frac{1}{4} - \frac{4}{4}} = \frac{-\frac{9}{2}}{-\frac{3}{4}} = \frac{9}{2} \times \frac{4}{3} = \frac{36}{6} = 6$$

Choice **a** is incorrect. When plugging in the given value of x, the 5 is subtracted, not multiplied.

Choice **b** is incorrect. When simplifying a fraction over a fraction, the fraction in the numerator is multiplied by the reciprocal of the fraction in the denominator. Dividing the fractions piece by piece is not a valid method.

Choice **d** is incorrect. Taking a value to the power of 2 is not the same as multiplying it by 2. Furthermore, a fraction with a denominator of zero is undefined, not equal to zero.

52. The correct answer is choice b. Subtract the two quantities:

$$
\begin{aligned}
(2.9 \times 10^9) - (3.2 \times 10^8) &= (29 \times 10^8) - (3.2 \times 10^8) \\
&= (29 - 3.2) \times 10^8 \\
&= 25.8 \times 10^8 \\
&= 2.58 \times 10^9
\end{aligned}
$$

Choice **a** is incorrect because you subtracted incorrectly. You must first convert the quantities to those that involve the same power of 10; then you can subtract the decimal parts. Choice **c** is incorrect because you cannot subtract $3.2 - 2.9$ and keep 10^9 as the power of 10 like this. Convert (2.9×10^9) to (29×10^8) and then subtract the decimal parts. Choice **d** is incorrect because it is the result of adding instead of subtracting the quantities.

53. The correct answer is choice b. Convert the radicals to fractional exponents and apply the exponent rules:

$$\frac{\sqrt{x}}{x \cdot \sqrt[3]{x}} = \frac{x^{\frac{1}{2}}}{x \cdot x^{\frac{1}{3}}} = \frac{x^{\frac{1}{2}}}{x^{\frac{4}{3}}} = x^{\frac{1}{2} - \frac{4}{3}} = x^{-\frac{5}{6}}$$

Choice **a** is incorrect because $\frac{x^a}{x^b} = x^{a-b}$, not x^{a+b}.

Choice **c** is incorrect because $\sqrt[m]{x^n} = x^{\frac{n}{m}}$ not $x^{\frac{m}{n}}$.

Choice **d** is incorrect because $\frac{x^a}{x^b} = x^{a-b}$, not $x^{\frac{a}{b}}$.

54. The correct answer is choice d. Apply the exponent rules, as follows:

$$\frac{2^3 \times 7^3}{14^{-2}} = \frac{(2 \times 7)^3}{14^{-2}} = \frac{14^3}{14^{-2}} = 14^3 \cdot 14^2 = 14^5$$

Choice **a** is incorrect because the sign of the exponent is incorrect. Choice **b** is incorrect because $\frac{x^b}{x^c} = x^{b-c}$, not x^{b+c}. Choice **c** is incorrect because $x^b \cdot y^b = (x \cdot y)^b$, not $(x \cdot y)^{b+b}$.

55. **The correct answer is choice c.** A number can be written in scientific notation as the product of a number between 1 and 10, including 1 and excluding 10, and a power of 10. If a number is less than 1, the power of 10 is negative. So, to write 0.00231 in scientific notation, start at the decimal point and move it to the right until you have one non-zero digit to the left of it. You must move the decimal point three places to get 2.31, so you multiply that number by 10 to the power of −3 because the number 0.00231 is smaller than 1 and you moved the decimal point three places to the right. So, in scientific notation:

$$0.00231 = 2.31 \times 10^{-3}$$

Choice **a** is incorrect since this product equals 0.231 (the exponent of −3 moves the decimal point to the left three times).

Choice **b** is incorrect since although this product equals 0.00231, the first number, 231, is not greater than 1 but less than 10.

Choice **d** is incorrect since this is the scientific notation for 2,310 since the exponent of 3 moves the decimal point to the right three times.

56. **The correct answer is choice d.** The distributive property pulls the greatest common factor out of a sum or difference of terms and represents it as the product of the greatest common factor multiplied by the factors of the sum or difference. For example, $Ax + Ay$ is represented as $A(x + y)$ through the distributive property. Since 8 is the greatest common factor of 64 and $8x$, the expression $64 - 8x$ can be written as $8(8 - x)$ or $(8 - x)8$.

Choice **a** is incorrect because 4 is not the greatest common factor of 64 and $8x$, so this product is not in lowest terms.

Choices **b** and **c** are incorrect because they do not demonstrate the distributive property.

57. **The correct answer is choice c.** Both of these numbers, 20 and 12, have a factor of 4, which is a perfect square. Therefore, these square roots can be simplified.

$$\sqrt{20} = \sqrt{4 \cdot 5} = \sqrt{4} \cdot \sqrt{5} = 2\sqrt{5}$$

$$\sqrt{12} = \sqrt{4 \cdot 3} = \sqrt{4} \cdot \sqrt{3} = 2\sqrt{3}$$

So, $\sqrt{20} \cdot \sqrt{12} = 2\sqrt{5} \cdot 2\sqrt{3}$. When simplified further, the answer is $4\sqrt{15}$.

Choice **a** is incorrect since 10 and 6 are *half* of 20 and 12, respectively, but they are not the *square roots* of 20 and 12.

Choice **b** is incorrect since although $\sqrt{240} = \sqrt{20} \cdot \sqrt{12}$, it should have been simplified to $4\sqrt{15}$ since it is not in lowest terms.

Choice **d** is incorrect since when you pulled out the perfect square of 4 as a factor of both 20 and 12, you forgot to take the square root of each of these 4's and instead multiplied them together.

58. **The correct answer is choice a.** Wherever we see an x, we need to replace it with the given value of -3. After substituting, the equation reads:

$$2(-3)^2 + -3 - 4$$

Remember, PEMDAS. Exponents come first. Which number is being squared? The parentheses indicate that it is just the -3. What is $(-3)^2$? $(-3)(-3) = +9$.

$$2(9) + -3 - 4$$

Multiplication comes next in the order of operations. What is being multiplied? $2(9) = 18$.

$$18 + -3 - 4$$

We have only addition and subtraction left, so we now just solve the problem left to right.

$$18 + -3 - 4$$
$$15 - 4 = 11$$

Choice **b** is incorrect because you substituted $+3$, and not -3, into the expression.

Choice **c** is incorrect because in the part of expression that reads $2(-3)^2$, you should have done $(-3)^2$ first, but instead you incorrectly multiplied 2 by -3 and then squared -6.

Choice **d** is incorrect because when you evaluated $2(-3)^2$, you got -18, but the negative should have canceled out when you squared -3, and this first part of the expression should have equaled 18.

59. The correct answer is 187. Evaluate $-3 + -10(-5 - 14)$ by first performing the subtraction in the parentheses, then multiplying that difference by -10, and last by adding that product to -3:

$$-3 + -10(-5 - 14) = -3 + -10(-5 - 14) = -3 + -10(-19)$$
$$= -3 + 190 = 187$$

60. The correct answer is choice d. Evaluate by first squaring -5 and cubing -4. The square of a negative number is always positive, and the cube of a negative number is always negative:

$$(-5)^2 - (-4)^3 = 25 - (-64)$$

This simplifies to $25 + 64 = 89$.

Choice **a** is incorrect because 2 is the solution to $(-5) \times 2 - (-4) \times 3$ and not the solution to $(-5)^2 - (-4)^3$. Remember that exponents don't represent simple multiplication, but repeated multiplication.

Choice **b** is incorrect because -39 is the solution to $(-5)^2 - (4)^3$ and not the solution to $(-5)^2 - (-4)^3$.

Choice **c** is incorrect because 39 is the solution to $-5^2 - (-4)^3$ and not the solution to $(-5)^2 - (-4)^3$. You forgot that the cube of a negative number will still be a negative.

61. The correct answer is choice a. In order to simplify this expression, first perform the multiplication in the numerator and the division in the denominator before doing any of the addition or subtraction:

$$\frac{4 + 3 \times 2}{7 - 14 \div 7} = \frac{4 + 6}{7 - 2}$$

Next, find the sum in your numerator and the difference in your denominator:

$$\frac{4 + 6}{7 - 2} = \frac{10}{5} = 2$$

Choice **b** is incorrect because although you performed the multiplication before the addition correctly in the denominator, you did not follow the correct order of operations in the numerator.

Choice **c** is incorrect because you performed the addition and subtraction in the numerator and denominator before doing the multiplication and division, and this is not the correct order of operations. You also dropped a negative sign when doing $7 - 14$.

Choice **d** is incorrect because you performed the addition and subtraction in the numerator and denominator before doing the multiplication and division, and this is not the correct order of operations.

62. **The correct answer is choice c.** In order to write 93,000,000 in scientific notation, it must be written in the form $a \times 10^b$ where $1 \leq a < 10$ and b is an integer. First move the decimal to define a as 9.3. Then b must be 7 in order to move the decimal in 9.3 to the right 7 times in order to arrive at 93,000,000: 9.3×10^7.

Choices **a**, **b**, and **d** are incorrect because the first number in scientific notation must be greater than or equal to 1, but less than 10.

63. **The correct answer is –25.** Begin by taking care of the first absolute value bracket: $|{-20} + 5| - |{-40}| = |{-15}| - |{-40}|$. Then take the absolute value of both numbers and subtract them to find the final answer: $15 - 40 = -25$.

64. **The correct answer is choice b.** First, substitute –2 in for both x's and rewrite the expression as: $4(-2)^2 + 3(1 - (-2))$ The order of operations indicates we must take care of parentheses first:

$$4(-2)^2 + 3(1 - (-2)) = 4(-2)^2 + 3(3)$$

We cannot multiply 4×-2 and then square the product because exponents come before multiplication in the order of operations. Take care of the exponents before multiplying:

$$4(-2)^2 + 3(3) = 4(4) + 3(3)$$

Last, perform the multiplication, followed by addition:

$$4(4) + 3(3) = 16 + 9 = 25$$

Choice **a** is incorrect because you didn't do the parentheses first. Instead you multiplied 3×1 only, and then subtracted the –2 to get 5 for the second half of the equation.

Choice **c** is incorrect because when calculating $4(-2)^2$, you multiplied 4×-2 before squaring it.

Choice **d** is incorrect because you multiplied 4×-2 before squaring it, as well as making the same mistake of omitting the parentheses that was outlined in choice **a**.

65. The correct answer is choice a. First, evaluate each of the expressions. Remember: A negative exponent means raise the reciprocal of the number to the opposite of that power (or flip it):

$$(\tfrac{2}{3})^{-2} = (\tfrac{3}{2})^{2} = \tfrac{3^2}{2^2} = \tfrac{9}{4}$$
$$(-3)^2 = (-3)(-3) = 9$$
$$(16)^{-1} = \tfrac{1}{16}$$
$$(\sqrt{5})^4 = (\sqrt{5})(\sqrt{5})(\sqrt{5})(\sqrt{5}) = \sqrt{625} = 25$$
$$(-2)^3 = (-2)(-2)(-2) = -8$$

Choose the greatest and the least of the values and find their sum by adding. Since 25 is the greatest value and –8 is the least value, the sum is $25 + -8 = 17$.

Choices **b**, **c**, and **d** are incorrect because each is the sum of two values in the set that are NOT the least and the greatest of the possible choices.

66. The correct answer is choice a. Substitute –3 for every t and rewrite the expression:

$$\tfrac{(t+6)-5}{t^2+t-7} = \tfrac{((-3)+6)-5}{(-3)^2+(-3)-7} = \tfrac{(3)-5}{9-3-7} = \tfrac{-2}{-1} = 2$$

Choices **b** and **d** are incorrect; they result from not correctly simplifying the negative signs.

Choice **c** is incorrect because it results when +3, not –3, is substituted for t.

67. The correct answer is choice d. Order of operations says that you should first simplify any expression in parentheses, then perform all multiplication and division, moving from left to right, and then perform all addition and subtraction, again moving from left to right. Therefore, to simplify this expression, start inside the parentheses and perform the multiplication and the division in the order in which they occur. Then, add the product and the quotient inside the parentheses to get 14, and finally, multiply 14 by 2.

$$2 \times (3 \times 4 + 6 \div 3) = 2 \times (12 + 6 \div 3) = 2 \times (12 + 2) = 2 \times 14 = 28$$

Choice **a** is incorrect because it reflects the result of evaluating the expression as follows: $(2 \times 3 \times 4 + 6) \div 3$.

Choice **b** is incorrect because you may have evaluated the expression as follows: $2 \times (3 \times 4 + 6) \div 3$.

Choice **c** is incorrect because you interpreted the expression as $2 \times 3 \times (4 + 6) \div 3$.

68. The correct answer is choice a. The expression $(-2)^5$ stands for negative 2 used as a factor 5 times:

$$-2 \times -2 \times -2 \times -2 \times -2 = -32$$

Remember, a negative nonzero number raised to an odd power is always negative.

Choice **b** is incorrect because it represents $-(5^2)$, or $-1 \times 5 \times 5 = -25$, which is not the same as $(-2)^5$.

Choice **c** is incorrect because it represents $(-5)2$, or $-5 \times -5 = 25$, which is not the same as $(-2)^5$.

Choice **d** is incorrect because it represents 25, which is $2 \times 2 \times 2 \times 2 \times 2$.

69. The correct answer is choice c. When multiplying factors that have like bases, such as 10^3 and 10^4, you can add the exponents. So, to simplify this expression, first use the commutative law of multiplication and rearrange the factors. Then, simplify each part of the expression:

$$(5 \times 10^3) \times (3 \times 10^4) = (5 \times 3) \times (10^3 \times 10^4)$$
$$= 15 \times 10^{3+4}$$
$$= 15 \times 10^7$$
$$= 1.5 \times 10^8$$

Choice **a** is incorrect because the factors 3 and 5 were added rather than multiplied.

Choice **b** is incorrect because the factors 3 and 5 were added rather than multiplied, and the powers of 10 were multiplied rather than added.

Choice **d** is incorrect because the powers of 10 were multiplied rather than added.

70. The correct answer is choice c. The $\sqrt{108}$ can be rewritten as $\sqrt{12 \cdot 9}$. The $\sqrt{32}$ can be rewritten as $\sqrt{8 \cdot 4}$. The problem now reads $\sqrt{12 \cdot 9} \cdot \sqrt{8 \cdot 4}$. This can be simplified further because 12 and 8 can be written as a product of two factors that contain perfect squares: $\sqrt{4 \cdot 3 \cdot 9} \cdot \sqrt{4 \cdot 2 \cdot 4}$. The square roots of the perfect squares, 4 and 9, can be simplified and pulled out of the roots: $2 \cdot 3\sqrt{3} \cdot 2 \cdot 2\sqrt{2}$. Multiply the whole numbers together: $2 \cdot 3 \cdot 2 \cdot 2 = 24$. Multiply the roots together: $\sqrt{3} \cdot \sqrt{2} = \sqrt{6}$. The complete answer is $24\sqrt{6}$.

71. The correct answer is choice b. The order of operations, PEMDAS, says that multiplication comes before subtraction. Thus, all of the first five terms in 1, 2, and 3 are multiplied *before* subtracting 2, or in the case of item 3, a quantity of 2.

Choice **a** is incorrect because it incorrectly groups the first four terms together and the last two terms together. All multiplication comes first.

Choice **c** is incorrect because it does not include the subtraction of 2.

Choice **d** is incorrect because it puts the terms in the incorrect order.

72. The correct answer is choice b. This expression equals $(\frac{1}{6}) \cdot (\frac{2}{3})^4 = \frac{1}{6} \cdot \frac{16}{81} = \frac{8}{243}$, which is a rational number.

Choice **a** is incorrect because while this is written as a fraction, the numerator and denominator are not integers.

Choice **c** is incorrect because this product equals $2^{\frac{1}{3}} \cdot 2^{\frac{1}{2}} = 2^{\frac{5}{8}}$, which is irrational.

Choice **d** is incorrect because this equals $9\pi^2$, which is a product of a rational number and an irrational number and so must be irrational.

73. The correct answer is choice d. Use the exponent rules to simplify as follows:

$$\frac{x^{\frac{3}{2}}}{x^{\frac{1}{4}}} = x^{\frac{3}{2} - \frac{1}{4}} = x^{\frac{5}{4}}$$

So, $\frac{5}{4}$ should be inserted in the box.

Choice **a** is incorrect; $\frac{x^a}{x^b} = x^{a-b}$, not $x^{\frac{a}{b}}$.

Choice **b** is incorrect because $\frac{x^a}{x^b} = x^{a-b}$, not x^{a+b}, and $\sqrt[m]{x^n} = x^{\frac{n}{m}}$, not $x^{\frac{m}{n}}$.

Choice **c** is incorrect because $\sqrt[m]{x^n} = x^{\frac{n}{m}}$, not $x^{\frac{m}{n}}$.

74. The correct answer is choice b. Apply the distributive property and then combine terms like this:

$$3x(1 - 2x) + 4x^2(x - 3) - 2x^3$$
$$= 3x - 6x^2 + 4x^3 - 12x^2 - 2x^3$$
$$= 2x^3 - 18x^2 + 3x$$

Choice **a** is incorrect because you only multiplied the first term of each binomial by the expression outside; you must use the distributive property to correctly perform this multiplication. You also ignored the $-2x^3$ term at the end.

Choice **c** is incorrect because you forgot to distribute the negative sign in the first binomial when multiplying $3x \times 2x$.

Choice **d** is incorrect because the sign of $18x^2$ is incorrect and the first and last terms are incorrect.

75. The correct answer is choice d. First, convert the quantities to ones involving the same power of 10. Note that $3.1 \times 10^5 = 31 \times 10^4$. So the difference is $(31 \times 10^4) - (2.3 \times 10^4) = (31 - 2.3) \times 10^4 = 28.7 \times 10^4 = 2.87 \times 10^5$.

Choice **a** is incorrect. You added instead of subtracting and did so incorrectly; make certain that when adding or subtracting quantities expressed using scientific notation that you first represent them all using a common power of 10.

Choice **b** is incorrect because you subtracted incorrectly. When adding or subtracting quantities expressed using scientific notation, you first represent them all using a common power of 10, and you cannot simply subtract exponents.

Choice **c** is incorrect because you added instead of subtracting.

76. The correct answer is choice b. Use the exponent rules to simplify, as follows:

$$2^4 \times 25^2 = 2^4 \times (5^2)^2 = 2^4 \times 5^4 = (2 \times 5)^4$$

Choice **a** is incorrect because you cannot combine the terms like this because the bases and exponents are both different in the given form.

Choice **c** is incorrect because the bases are different so you cannot combine the terms by simply adding the exponents.

Choice **d** is incorrect because $a^b \cdot c^b = (a \cdot c)^b$, not $a^b \cdot c^b = (a \cdot c)^{b+b}$.

77. The correct answer is choice b. You can easily compare this option with choice **c**, since they are both expressed in terms of $(\frac{2}{3})^t$, and see that the two options cannot be the same ($\frac{171}{2} \neq 38$). This option is the only one that cannot be transformed to look exactly like choice **c**. Choice **a** is incorrect because if you change this equation so that it's expressed in terms of $(\frac{2}{3})^t$, you get $C = 57(\frac{2}{3})^{t+1} = 57 \cdot (\frac{2}{3}) \cdot (\frac{2}{3})^t = 38(\frac{2}{3})^t$, which is the same as choice **c**.

Choice **d** can be transformed in the same way.

Choice **c** is incorrect because when choices **a** and **d** are transformed so that they are also expressed in terms of $(\frac{2}{3})^t$, they look just like this option and are therefore equivalent to it. Only choice **b** cannot be made to look just like this option.

Choice **d** is incorrect because if you change this equation so that it's expressed in terms of $(\frac{2}{3})^t$, you get $C = 38(\frac{3}{2})^{-t} = 38[(\frac{3}{2})^{-1}]^t = 38(\frac{2}{3})^t$, which is the same as choice **c**.

Choice **a** can be transformed in the same way.

78. The correct answer is choice b. Convert the radicals to rational exponents and apply the exponent rules:
$$\sqrt[3]{x^2} \cdot \sqrt{x^5} = x^{\frac{2}{3}} \cdot x^{\frac{5}{2}} = x^{\frac{2}{3} + \frac{5}{2}} = x^{\frac{19}{6}}$$
Choice **a** is incorrect because $x^a \cdot x^b = x^{a+b}$, not $x^{a \cdot b}$.
Choice **c** is incorrect because $\sqrt[m]{x^n} = x^{\frac{n}{m}}$, not $x^{\frac{m}{n}}$.
Choice **d** is incorrect because $x^a \cdot x^b = x^{a+b}$, not $x^{a \cdot b}$, and $\sqrt[m]{x^n} = x^{\frac{n}{m}}$, not $x^{\frac{m}{n}}$.

79. The correct answer is choice d. First, we need to determine the number of seconds in one day: 1 minute = 60 seconds, 1 hour = 60 minutes = 60(60) = 3,600 seconds, so 1 day = 24 hours = 24(3,600) seconds = 86,400 seconds. Therefore, light travels $(86,400) \times (3.2 \times 10^8) = 2.7648 \times 10^{13}$ meters in a day.

Choice **a** is incorrect because you multiplied by the number of hours in a day, but you did not multiply by the number of seconds in an hour.

Choice **b** is incorrect because this is the distance traveled in 1 minute, not 1 day.

Choice **c** is incorrect because this is the distance traveled in 1 week, not 1 day.

80. **The correct answer is choice d.** Convert the radicals to rational exponents and use the exponent rules to simplify:

$$\sqrt[3]{\sqrt{\tfrac{1}{5}}} = \left(\left(5^{-1}\right)^{\frac{1}{2}}\right)^{\frac{1}{3}} = 5^{(-1) \cdot \left(\frac{1}{2}\right) \cdot \left(\frac{1}{3}\right)} = 5^{-\frac{1}{6}}$$

Choice **a** is incorrect because the exponent has the wrong sign. $\frac{1}{5} = 5^{-1}$.

Choice **b** is incorrect because you added the exponents instead of multiplying them. $(a^b)^c = a^{b \cdot c}$.

Choice **c** is incorrect because $\sqrt[m]{a} = a^{\frac{1}{m}}$, not a^m.

3

Rates, Proportions, and Percents

Rates, proportions, and percents are concepts that we encounter all around us: miles per hour, price per pound, percentage discounts, and commissions. In this chapter you will be tested on some extremely useful skills that can be applied to your personal and work life. In addition to being useful in your everyday life, mastery of the following skills that are tested in this chapter will be critical to your success on the GED test:

- Rates and unit rate

- Setting up and solving proportions

- Working with percentages

- Solving word problems with percentages

81. Suppose a jet can fly a distance of 5,100 miles in 3 hours. If the jet travels at the same average speed throughout its flight, how many hours will it take the jet to travel 22,950 miles?
    a. 4.5 hours
    b. 13.5 hours
    c. 15.3 hours
    d. 18 hours

82. Joe made $90 babysitting for 12 hours. At this rate, how long will it take him to make an additional $300?
    a. 25 hours
    b. 7.5 hours
    c. 40 hours
    d. 28 hours

83. Jeremy purchased six cans of tomatoes for $5.34. At this rate, how much would he pay for 11 cans of tomatoes?
    a. $10.68
    b. $9.79
    c. $9.90
    d. $11.00

84. Solve for x: $\frac{8}{10} = \frac{x}{100}$. _____

85. The scale on a state map is 1 inch:24 miles. How many miles apart are two cities if they are 3 inches apart on the map?
    a. 32 miles
    b. 72 miles
    c. 80 miles
    d. 96 miles

86. A survey of 1,000 registered voters shows that 650 of them will choose Candidate A in an upcoming election. If 240,000 people vote in the upcoming election, according to the survey, how many votes will Candidate A receive? _____

87. Samantha went to a local restaurant to celebrate her birthday with a friend. The charge for the meal was $15. Samantha paid with a $20 bill and tipped the waiter 15% of the cost of the meal. How much change did she have left?
   a. $2.25
   b. $2.75
   c. $3.50
   d. $3.75

88. If Veronica deposits $5,000 in her savings account with a yearly interest rate of 9% and leaves the money in the account for eight years, how much interest will her money earn?
   a. $360,000
   b. $45,000
   c. $3,600
   d. $450

89. Mr. Jordan is planning to buy a treadmill. The treadmill he wants is on sale at 10% off the retail price of $700. Mr. Jordan has an additional coupon for 5% off after the discount has been applied. What is the final cost of the treadmill, not including any taxes or assembly fees?
   a. $595.00
   b. $598.50
   c. $630.00
   d. $661.50

90. Alexis bought a gardening shed for $339. She loved it so much that the next summer she went to buy another one, but the price had gone up to $419. What was the percentage increase in price of her beloved garden shed, rounded to the nearest whole number?
   a. 24%
   b. 36%
   c. 19%
   d. 80%

91. A website is selling a laptop computer for $375 plus 6.5% state sales tax. A student wishes to purchase two of these computers: one for his brother and one for himself. Including tax, what will be the total cost of his order? Write your answer in the box.

$ [＿＿＿＿＿]

92. A town has a population of 20,510 and an area of 86.8 square miles. To the nearest tenth, what is the population density as measured by people per square mile?
a. 2.72
b. 236.3
c. 2,201.4
d. 55,833.1

93. Two friends go to a restaurant for lunch and receive a final bill of $24.36. One friend believes they should tip 15%, while the other believes they should tip 20%. To the nearest cent, what is the difference between the two possible tips?
a. $1.22
b. $3.65
c. $4.87
d. $8.52

94. A map is drawn such that 2.5 inches on the map represents a true distance of 10 miles. If two cities are 7.1 inches apart on the map, then to the nearest tenth of a mile, what is the true distance between the two cities?
a. 14.6
b. 17.8
c. 28.4
d. 71.0

95. Over the last 6 months, a company's monthly revenue has increased by 28%. If the revenue this month is $246,990, then what was the revenue 6 months ago? Round your answer to the nearest cent. Write your answer in the box.

$ [＿＿＿＿＿]

96. A remote-controlled vehicle travels at a constant speed around a testing track for a period of 12 hours. In those 12 hours, the vehicle covers 156 kilometers. In terms of kilometers per hour, at what rate was the vehicle traveling? Write your answer in the box.

| | km/hr.

97. An IT consultant charges a company $75 an hour to analyze its current systems. Additionally, he charges a 3% project fee and a 1% telecommunications fee on the cost of the billed hours. If a project requires 20 hours for the consultant to complete, what will be the final amount charged to the company?
  a. $1,515
  b. $1,545
  c. $1,560
  d. $2,100

98. The ratio of full-time employees to part-time employees in a mid-size law firm is 4:3. If there is a total of 20 full-time employees, how many part-time employees work at the firm?
  a. 15
  b. 19
  c. 23
  d. 27

99. A company pays its sales employees a base rate of $450 a week plus a 4% commission on any sales the employee makes. If an employee makes $1,020 in sales one week, what will be his total paycheck for that week? Write your answer in the box.

100. A 32-ounce bag of potato chips has a retail cost of $3.45. To the nearest 10th of a cent, what is the price per ounce of this item (in cents)?
  a. 9.3
  b. 10.8
  c. 28.5
  d. 35.45

101. A walking trail is 11,088 feet long. If a mile is 5,280 feet, how many miles long is the walking trail?

a. 0.2

b. 0.5

c. 1.6

d. 2.1

102. The following table indicates the behavior of the price of one share of a given stock over several weeks.

END OF	CHANGE
Week 1	Increased by $5.00
Week 2	Decreased by 10%
Week 3	Decreased by $1.10
Week 4	Doubled in value

If the stock was worth $10.15 a share at the beginning of week 1, what was the value of one share of this stock at the end of week 4?

a. $25.07

b. $29.46

c. $32.20

d. $50.14

103. A customer uses two coupons to purchase a product at a grocery store, where the original price of the product was $8.30. If the final price paid by the customer was $7.00 and each coupon gave the same discount, what was the value of the discount provided by a single coupon?

a. $0.65

b. $0.90

c. $1.30

d. $2.60

104. Lee is planning to buy a new television and has been watching the price of a particular model for the past month. Last month, the price was $309.99, while this month, the price is $334.99. To the nearest tenth of a percent, by what percent has the price increased over the past month? Write your answer in the box.

%

105. Ava bought 2.5 pounds of organic green grapes at the farmers' market for $9. Liam bought 3.2 pounds of organic red grapes at the store for $12. To the nearest penny, how much more per pound were the grapes that Liam bought at the store?
   a. $3.00
   b. $3.68
   c. $0.15
   d. $0.25

106. Nine out of ten professional athletes suffer at least one injury each season. If there are 120 players in a professional league, estimate how many of them played the entire last season without an injury.
   a. 12
   b. 13
   c. 100
   d. 108

107. Calculate: 320 is 40% of what number?
   a. 128
   b 360
   c. 780
   d. 800

108. Lucas is buying a drill that costs $120, but on July 4th it is on a one-day sale for 30% off. If the sales tax is 7.5%, what will the total price be after calculating the 30% discount and tax? _____

109. If K.P. can read 1,000 words in 5 minutes, how many words can he read in 12 minutes? _____

110. The ratio of men to women at a certain meeting is 3 to 5. If there are 18 men at the meeting, how many people are at the meeting?
   a. 11
   b 29
   c. 30
   d. 48

111. A drawing of a bookcase to be built has a scale of 1 inch:1 foot. The height of the bookcase in the drawing is 2.5 inches. What is the actual height in inches of the bookcase?

    a. 12 inches

    b 18 inches

    c. 24 inches

    d. 30 inches

112. Suzette just moved to a new city. She does not have a car, so she has to walk or take the bus everywhere. She bought a map of the city so she can find her way around. She located a grocery store on the map, and it doesn't appear to be too far from her apartment. From the scale drawing, determine how many miles away the grocery store is from Suzette's apartment. *Note:* 1 mile = 5,280 feet.

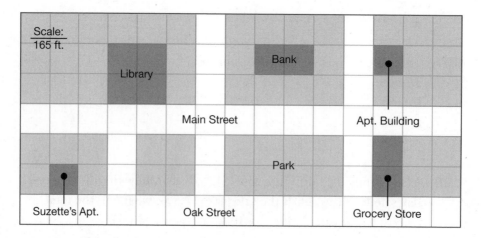

    a. 0.2 miles

    b 0.3 miles

    c. 6.4 miles

    d. 825 miles

113. A furniture store manager decides to mark up the price of a living room set by $q\%$ in anticipation of increased interest during the holiday gift-buying season. If the new price is \$z before tax, what was the price of the set before the markup?

a. $\dfrac{100}{100 + q}z$

b. $\dfrac{100 + q}{100}z$

c. $z + q$

d. $\dfrac{100}{q}z$

114. Ralph can hike 1.3 miles in 45 minutes. Which equation could be used to find d, the distance in miles that Ralph can hike in 3 hours?

a. $\dfrac{d}{3} = \dfrac{0.75}{1.3}$

b. $\dfrac{1.3}{0.75} = \dfrac{d}{3}$

c. $\dfrac{0.75}{d} = \dfrac{3}{1.3}$

d. $\dfrac{0.75}{3} = \dfrac{d}{1.3}$

115. The following graph shows students' favorite sports. What percentage said that golf was their favorite?

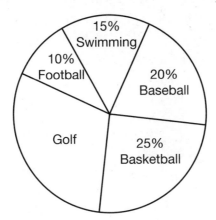

a. 25%

b. 30%

c. 45%

d. 15%

116. A certain model of calculator is known to have approximately 8 malfunctioning calculators out of every 2,000. If an office supply company in Los Angeles is going to order 750 of these calculators to stock its stores for back-to-school shopping in August, approximately how many returns due to defects should the company anticipate having later in the fall? _____

117. It takes Michael 3 minutes to read two pages in his book. At this rate, how long will it take him to read a 268-page book?
   a. 6 hours and 42 minutes
   b 13 hours and 24 minutes
   c. 8 hours and 56 minutes
   d. 5 hours and 2 minutes

118. A drawing of a tree uses the scale 1 in. = 25 ft. If the drawing is 1.5 inches tall, how tall is the tree shown in the drawing?

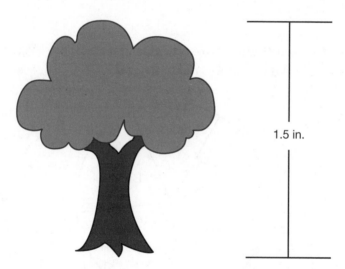

1.5 in.

   a. 26 ft.
   b 37.5 ft.
   c. 50 ft.
   d. 62.5 ft.

119. Ron is a car salesman. If he meets his daily sales goal, he gets 5% commission on the amount of money he was over his goal. If Ron sold a Jeep Cherokee and a Ford Escape, how much commission did he get if his daily sales goal was $42,300?

CAR	PRICE
Ford Escape	$22,700
Honda SUV	$22,795
Kia Sorento	$24,100
Jeep Cherokee	$22,995
Chevrolet Equinox	$23,755
Dodge Journey	$19,485

   a. $1.69
   b $16.97
   c. $169.75
   d. $16,975

120. The Robb family wants to have the carpeting in their vacation home steam cleaned. They received a special offer in the mail advertising $3 for every 10 square feet of carpeting cleaned. If they don't want to spend more than $250, how many square feet of carpeting can they have cleaned with this offer? _____

121. The following drawing represents a deck that the Joneses are building. In the drawing, one-half of an inch represents 3 feet. What are the actual dimensions of the deck?

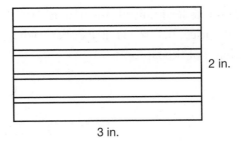

3 in.

   a. 3 feet by 4.5 feet
   b 6 feet by 9 feet
   c. 7 feet by 10.5 feet
   d. 12 feet by 18 feet

122. Arnita has been working for three years at the same company, and has been earning $11.75 per hour. She recently received a raise of $13 per hour. By what percentage has her earnings increased per hour?

   a. 0.096%

   b 0.106%

   c. 9.6%

   d. 10.6%

123. If $3\frac{3}{4}$ cups of flour are needed to bake a batch of 36 cookies, how much flour is needed to make 90 cookies?

   a. $5\frac{5}{8}$ cups

   b. $9\frac{3}{8}$ cups

   c. $8\frac{1}{2}$ cups

   d. $7\frac{1}{2}$ cups

124. A salad bar owner charges a base price of $5.75 for a basic salad with three toppings and $0.80 for each additional topping. A 6% tax is then applied to that price to get the total cost. Which of the following functions could be used to compute the cost of a salad based on the number of additional toppings x?

   a. $f(x) = (5.75 + 0.80x) + 6(5.75 + 0.80x)$

   b. $f(x) = 5.75 + 0.80x$

   c. $f(x) = 5.75 + 0.80x + 0.06$

   d. $f(x) = (5.75 + 0.80x) + 0.06(5.75 + 0.80x)$

125. An electronics retailer reduced the price of last year's laptops by p%. If the sale price, before tax, is x dollars, what was the original price (in dollars) of the laptop?

   a. $\frac{100}{100 + p}x$

   b. $\frac{100 - p}{100}$

   c. $\frac{x}{p}$

   d. $\frac{100x}{100 - p}$

126. Sandy's Treasures sells used books in bundles at a great discount. According to the table, how much would it cost to buy nine used books? _____

BUNDLES OF BOOKS	PRICE
3	$15
6	$30
9	x
12	$60
15	$75

127. If Sam jogs at a rate of 6 miles per hour, approximately how many minutes would it take him to jog 5,000 feet?
  a. 833
  b. 1,704.6
  c. 28.4
  d. 9.5

128. A dinner bill comes to $75.50 before applying the 6% tax. You have a coupon for 15% off the bill before tax and intend to tip 20% on the taxed amount after the coupon is applied. What is the total amount you will pay?
  a. $81.64
  b. $77.02
  c. $64.18
  d. $68.03

129. The number of cell phones in the United States in 2014 was 327,577,529. If 46% of these cell phones were owned by males, how many cell phones were owned by males? Round your answer to the nearest hundred thousand.
  a. 150,000,000
  b. 150,690,000
  c. 150,680,000
  d. 150,700,000

130. If the mass of five identical containers of soup is 3.2 kilograms, what would be the mass of eight such identical containers of soup?

    a. 12.5 kilograms

    b. 5.12 grams

    c. 6.4 kilograms

    d. 5,120 grams

Answers and Explanations

81. The correct answer is choice b. Dividing 5,100 miles by 3 hours gives you the speed of the jet in miles per hour:

$\frac{5,100}{3}$ = 1,700 miles per hour

Therefore, to fly a distance of 22,950 miles, divide the distance 22,950 miles by 1,700 miles per hour, which equals $\frac{22,950}{1,700}$ = 13.5 hours. Choice **a** is incorrect because 4.5 is the answer gotten when 22,950 miles is divided by 5,100 miles. However, since the jet flies 5,100 in 3 hours, you must first find the unit speed of miles per hour that the jet flies (5,100 ÷ 3 = 1,700) and then divide 22,950 by 1,700.

Choice **c** is incorrect because it is the correct answer with the tenths digit and the ones digit transposed.

Choice **d** is incorrect because, if the jet flies 5,100 miles in 3 hours, it would fly 6 times that distance in 18 hours, which would be 5,100 × 6 = 30,600 miles and not 22,950 miles.

82. The correct answer is choice c. Find how much Joe makes per hour:

$90 ÷ 12 = $7.50

Joe makes $7.50 per hour.

To find how many hours he will need to babysit to earn $300, divide $300 by $7.50:

$300 ÷ 7.50 = 40

It will take Joe 40 hours to earn an additional $300.

Choice **a** is seen as incorrect by using some simply estimating: If Joe earned $90 in 12 hours, then in 25 hours he would earn slightly more than twice $90 (since 25 is slightly more than twice 12). Two times $90 is $180, which is much lower than the additional $300 Joe wants to earn. You may have divided $300 by 12 hours, but this is not the correct way to handle unit rate problems.

Choice **b** is incorrect because 7.5 represents Joe's hourly rate since $90 divided by 12 hours is $7.50 per hour.

Choice **d** is incorrect because if Joe earns $90 for 12 hours of work, then he is earning $7.50 per hour. After 28 hours of work at this rate, he would earn only $210.

83. **The correct answer is choice b.** To find the cost of one can of tomatoes, divide the cost of six cans ($5.34) by 6:

$5.34 ÷ 6 = $0.89

Each can of tomatoes costs $0.89.

Next, to find the cost of 11 cans, multiply $0.89 by 11:

$0.89 × 11 = $9.79

The cost of 11 cans of tomatoes is $9.79.

Choice **a** is incorrect because $10.68 is twice the price of 6 cans, which cost $5.34. This means that $10.68 would be the price of 12 cans, and not 11 cans.

Choice **c** is incorrect because it is slightly higher than the actual cost of $9.79.

Choice **d** is incorrect because if 11 cans cost $11 that would mean that each can cost $1 and that is not the case since 6 cans cost $5.34.

84. **The correct answer is $x = 80$.** Solve the proportion $\frac{8}{10} = \frac{x}{100}$ by setting the cross products equal to one another:

$$8(100) = 10(x)$$
$$800 = 10x$$
$$\frac{800}{10} = x$$
$$x = 80$$

85. **The correct answer is choice b.** Because 1 inch on the map represents 24 miles, 3 inches on the map represent 3 × 24, or 72 miles.

Choice **a** is incorrect because since every inch represents 24 miles, then two inches represents 24 × 2 = 48 miles, which is already greater than this answer choice.

Choice **c** is incorrect because since each inch represents 24 miles, a distance of 80 miles would be 80 ÷ 24 = 3.3 inches.

Choice **d** is incorrect because since each inch represents 24 miles, a distance of 96 miles would be 96 ÷ 24 = 4 inches.

86. The correct answer is 156,000. Solve this word problem by setting a proportion that compares Candidate A voters to total voters. The given information tells us that the first ratio will be 650 Candidate A voters to 1,000 total voters:

$$\frac{\text{Candidate A voters}}{\text{total voters}} = \frac{650}{1,000}$$

Set this ratio equal to a ratio with 240,000 as the total voters and A as the Candidate A voters:

$$\frac{\text{Candidate A voters}}{\text{total voters}} = \frac{650}{1,000} = \frac{A}{240,000}$$

Now solve this using cross products:

$$650(240,000) = 1,000A$$
$$156,000,000 = 1,000A$$
$$A = 156,000$$

Therefore, it can be anticipated that Candidate A will receive 156,000 out of the 240,000 votes.

87. The correct answer is choice b. A 15% tip on a charge of $15 equals $2.25. Therefore, the total amount that Samantha paid was $15.00 + $2.25 = $17.25. The difference equals $20.00 − $17.25 = $2.75.

Choice **a** is incorrect because $2.25 is 15% of the $15 bill, so you calculated the tip she left and not the change she had remaining.

Choice **c** is incorrect because if you calculated that Samantha had $3.50 remaining, that means the tip she left would have been $1.50, which is just 10% of the $15 bill and not 15% of the bill.

Choice **d** is incorrect because if you calculated that Samantha had $3.75 remaining, that means that she would have left only a $1.25 tip, which would be less than 10% of her $15 bill.

88. **The correct answer is choice c.** In the formula $I = prt$, the amount of money deposited is called the principal, p. The interest rate per year is represented by r, and t represents the number of years. The interest rate must be written as a decimal. Here, $p = 5,000$, $r = 9\% = 0.09$, and $t = 8$. Substitute these numbers for the respective variables and multiply: $I = 5,000 \times 0.09 \times 8 = \$3,600$.

Choice **a** is incorrect because the 9% rate must be written as a decimal in order to calculate interest. Instead you performed $\$5,000 \times 9 \times 8$ years, and this is not how interest is calculated.

Choice **b** is incorrect because you represented 9% as 9 instead of as 0.09. $\$45,000$ is $\$5,000 \times 9$, and this is not how interest is calculated, since the rate must be written as a decimal: 0.09. Additionally, the 8 years need to be taken into account.

Choice **d** is incorrect because $\$450$ is 9% of $\$5,000$ for just 1 year, not for 8 years.

89. **The correct answer is choice b.** This question requires taking your time and making sure you do all the required steps. Do one step at a time to arrive at the correct answer.

First, find the sale price after the 10% has been deducted (remember, 10% is the same as 0.10):

$$\$700 - 0.10(\$700) = \$700 - \$70$$
$$= \$630$$

Now, apply the 5% coupon to the discounted price of $630. Remember, 5% is the same as 0.05. Be sure to subtract from $630, not from the original price of $700.

$$\$630 - 0.05(\$630) = \$630 - \$31.50$$
$$= \$598.50$$

Choice **a** is incorrect because you took 15% off the original price, but you may not combine the 10% sale and the 5% coupon at the same time since the coupon must be applied after the sale discount has been taken.

Choice **c** is incorrect because you calculated the price of the treadmill after the 10% sale discount, but you didn't calculate the additional savings from the 5% coupon.

Choice **d** is incorrect because you added 5% onto the sale price of $630 rather than subtracting 5%.

90. The correct answer is choice a. Use the formula for % of increase by subtracting the original price from the new price:

% of increase: $= \frac{\text{amount of increase}}{\text{original amount}} \times 100$

% of increase $= \frac{419 - 339}{339} \times 100 = 23.59\%$

So the percentage increase is rounded to 24%. Answer choice **c**, 19%, is what you get if you divided the amount of increase by the new price of $419, but instead you have to divide it by the original price. Answer choice **d**, 80%, is just the difference in price from last year to this year, without it being divided by the original price. Choice **b** can be ruled out as incorrect with some careful estimating since a 36% increase would have increased the shed by more than $\frac{1}{3}$ the price, or more than $110; however, it increased by just $80. Choice **c** is incorrect because you compared the increase of $80 to the new price and not the original price. You should have done $\frac{80}{339}$ and not $\frac{80}{419}$.

Choice **d** is incorrect because 80 is the dollar amount of the increase in price, but it is not the percentage increase.

91. The correct answer is $798.75. The student will spend $375 × 2 = $750 on the two computers and $750 × 0.065 = $48.75 on the tax: $750 + $48.75 = $798.75.

92. The correct answer is choice b. Dividing the number of people by the area yields 236.29 ≈ 236.3.

Choice **a** is incorrect. The term *square miles* does not imply that the 86.8 must be squared. It is instead a unit of measure for area.

Choice **c** is incorrect. Since the final result will be people per square mile, taking the square root before dividing is not a needed step.

Choice **d** is incorrect. Although the area is measured in square miles, the values of the population and the area do not need to be squared.

93. The correct answer is choice a. 0.2 × 24.36 – 0.15 × 24.36 = 1.22.

Choice **b** is incorrect. This represents a tip of 15%, not the difference between the two tips.

Choice **c** is incorrect. This represents a tip of 20%, not the difference between the two tips.

Choice **d** is incorrect. The *difference* refers to subtraction, not addition.

94. **The correct answer is choice c.** If x is the number of miles between the two cities, then $\frac{2.5 \text{ in.}}{10 \text{ mi}} = \frac{7.1 \text{ in.}}{x \text{ mi}}$. Cross multiply and solve the resulting equation:

$$2.5x = 71$$
$$x = \frac{71}{2.5} = 28.4$$

Choice **a** is incorrect. This is a proportional relationship, so subtraction does not apply in general.

Choice **b** is incorrect. The final result must be in miles, but multiplying two values that are measured in inches will yield a result in square inches.

Choice **d** is incorrect. This would be the number of miles if each inch represented 10 miles.

95. **The correct answer is $192,960.94.** If x represents the revenue 6 months ago, then the equation $1.28x = 246,990$ must be true. Dividing both sides by 1.28 yields $x = 192,960.94$.

96. **The correct answer is 13.**

$$\frac{156 \text{ km}}{12 \text{ hr.}} = \frac{\frac{156}{12} \text{ km}}{\frac{12}{12} \text{ hr.}} = \frac{13 \text{ km}}{1 \text{ hr.}}$$

97. **The correct answer is choice c.** $75 \times 20 = 1,500$, $0.01 \times 1,500 = 15$, and $0.03 \times 1,500 = 45$ for a total of $1,500 + 15 + 45 = 1,560$.

Choice **a** is incorrect. This includes only the telecommunications fee, but there is also a 3% project fee.

Choice **b** is incorrect. This includes only the project fee, but there is also a 1% telecommunications fee.

Choice **d** is incorrect because 3% of a total is found by multiplying by 0.03, not 0.3. Similarly, 1% is found by multiplying by 0.01 instead of 0.1.

98. **The correct answer is choice a.** To maintain the ratio, the fraction of full-time employees to part-time employees must be equivalent to $\frac{4}{3}$. The number of full-time employees can be found by multiplying 4 by 5; therefore, the number of part-time employees can be found by multiplying 3 by 5 to get 15.

Choice **b** is incorrect. Although the difference between 20 and 4 is 16, it can't be used to find the final answer. Ratios work with a common multiplier, not a common sum.

Choice **c** is incorrect. This will not maintain the ratio, since it did not use a common multiplier.

Choice **d** is incorrect. This would be approximately correct if 4:3 were the ratio of part-time employees to full-time employees instead of vice versa.

99. **The correct answer is $490.80.** The employee is paid a 4% commission on his sales of $1,020. Therefore, he will be paid 0.04 × $1,020 = $40.80 for the sales. This is on top of his regular pay of $450. Therefore, his total paycheck will be $450 + $40.80 = $490.80.

100. **The correct answer is choice b.** The price per ounce is found by dividing 3.45 by 32.

Choice **a** is incorrect. Dividing the number of ounces by the cost will give the number of ounces per cent.

Choice **c** is incorrect. Subtracting terms will not give an interpretable value.

Choice **d** is incorrect. Adding these two terms will not give an interpretable value.

101. **The correct answer is choice d.** The conversion given can be written as a ratio: 1 mile:5,280 feet. Use this to cancel out units: $11,088 \text{ ft} \times \frac{1}{5,280} = \frac{11,088}{5,280} = 2.1$.

Choice **a** is incorrect. There is no need to divide by 12 since the units are not in inches.

Choice **b** is incorrect. Dividing 5,280 by 11,088 leaves the units in terms of $\frac{1}{\text{miles}}$, which doesn't make sense.

Choice **c** is incorrect. Subtracting the two values will not give an interpretable value.

102. **The correct answer is choice a.** After increasing by $5.00, the share was worth $15.15. It then decreased in value by 10%, or by 0.1 × 15.15 = 1.515. Therefore, at the end of week 2, it was worth $15.15 – $1.515 = $13.635 a share. At the end of week 3, it was worth $13.635 – $1.10 = $12.535. Finally, it doubled in value and was worth 2 × $12.535 = $25.07 per share.

Choice **b** is incorrect. The stock decreased in value by $1.10 at the end of week 3. This represents subtraction in the problem.

Choice **c** is incorrect. A 10% decrease can be found by multiplying 0.9 and the current value. This answer comes from using 1% or 0.01 as the decrease.

Choice **d** is incorrect. To double means to multiply by 2 and not 4.

103. **The correct answer is choice a.** If x represents the discount provided by a single coupon, then $2x$ represents the combined discount provided by both. Given the prices before and after, the following equation can be written and solved:

$$8.3 - 2x = 7$$
$$-2x = -1.3$$
$$x = 0.65$$

Choice **b** is incorrect. If each coupon gave a 90-cent discount, the final price would have been $8.30 - $1.80 = $6.50.

Choice **c** is incorrect. This is the value of both coupons together.

Choice **d** is incorrect. This is twice, rather than half, the value of the two coupons together.

104. **The correct answer is 8.1%.** The percent increase can be found by finding the difference between the two prices and then dividing by the original price:

$$\frac{334.99 - 309.99}{309.99} = 0.0806$$

Multiplying by 100 to convert this to a percentage yields 8.06%. Rounded, this is 8.1%.

105. **The correct answer is choice c.** First determine the unit price of Ava's organic green grapes by dividing $9 by 2.5 pounds: $\frac{\$9}{2.5} = \$3.60$. Then determine the unit price for Liam's organic red grapes: $\frac{\$12}{3.2} = \$3.75$. Last, subtract the unit prices to show that Liam's grapes cost $0.15 more per pound.

Choice **a** is incorrect because $3.00 is not the difference in price *per pound* but is simply the difference between the two different totals that Ava and Liam spent on grapes. Unit price must be used.

Choice **b** is incorrect because this is the average price between the two types of grapes since you added $9 and $12 and then divided that sum by the total number of pounds of grapes, which was 5.7 pounds. You must find each unit price separately.

Choice **d** is incorrect because per pound, Liam's grapes cost only $0.15 more, and not $0.25 more.

106. **The correct answer is choice a.** Since nine out of ten professional athletes suffer at least one injury per season, that means that one out of ten, or 10%, of professional athletes play the entire season without an injury: 10% of 120 is 12.

Choice **b** is incorrect because you divided 120 by 9, but this is not the correct way to work with percentages. You should have changed *nine out of ten* to a percentage or set up a proportion comparing the number of injury-free players to total players.

Choice **c** is incorrect because if only one out of ten players goes without an injury each season, there is no way that 100 out of 120 could go without an injury.

Choice **d** is incorrect because 108 is the number of players who suffered at least one injury last season and not the number of players who went without an injury.

107. **The correct answer is choice d.** When working with percentages, use the proportion $\frac{part}{whole} = \frac{\%}{100}$ to solve for the missing piece of information. In this case the percentage is 40, and 320 is the *part* since it is *of* an unknown number:

$$\frac{part}{whole} = \frac{\%}{100}$$
$$\frac{320}{x} = \frac{40}{100}$$
$$40x = 32,000$$
$$x = 800$$

Choice **a** is incorrect because you multiplied 40% by 320, but this gives the answer to a different question, "What is 40% *of* 320?"

Choice **b** is incorrect because you simply added 40 from the 40% to the 320, which is not the correct way to approach percentage problems.

Choice **c** is incorrect because 320 is 41% of 780 and not 40%.

108. **The correct answer is $90.30.** First, determine the sale price. Since the drill is 30% off, that means Lucas will pay 70% of the drill's normal price. 70% of 120 is 0.70 × $120 = $84. Next, determine the sales tax by taking 7.5% of $84: 0.075 × 84 = $6.30. Adding the tax onto the sale price of the drill will yield the total price of $90.30.

109. **The correct answer is 2,400.** Since K.P. can read 1,000 words in 5 minutes and you need to determine how many words can he read in 12 minutes, set up a proportion that models the pairs of information using the ratio $\frac{words}{minutes}$: $\frac{1,000}{5} = \frac{w}{12}$. Next, to solve for w, cross multiply: 1,000 × 12 = 5w, and then isolate w: $w = \frac{12,000}{5} = 2,400$.

110. **The correct answer is choice d.** The first step of the solution is to use the ratio $\frac{men}{women}$ to record the given information of there being 3 men and 5 women: $\frac{men}{women} = \frac{3}{5}$. Next, we want to turn this relationship into a proportion where the second ratio shows 18 men. Since we don't know the number of women at the meeting, we used w to represent them: $\frac{men}{women} = \frac{3}{5} = \frac{18}{w}$. Solve this proportion using equivalent cross products: $3w = 5(18)$. Divide both sides by 3 to get $w = 30$. This is not our answer to the question, though, which asked for the *total* number of men and women. Since there are 30 men and 18 women, there are 48 people in total at the meeting.

Choice **a** is incorrect because you set up your ratio upside down. It should have been $\frac{women}{men} : \frac{5}{3} = \frac{x}{18}$; however, you set it up as $\frac{women}{men} : \frac{5}{3} = \frac{18}{x}$ and then you forgot to add the 11 you got for x to the 18 men who were at the meeting.

Choice **b** is incorrect because you set up your ratio upside down. It should have been $\frac{women}{men} : \frac{5}{3} = \frac{x}{18}$; however, you set it up as $\frac{women}{men} : \frac{5}{3} = \frac{18}{x}$ and then you added that incorrect number of women (11) to the 18 men who were at the meeting.

Choice **c** is incorrect because 30 is the number of women at the meeting and is *not* the total number of people at the meeting. You forgot to add the 30 women to the 18 men to get a total of 48 people.

111. **The correct answer is choice d.** One inch represents 1 foot, so you can use a proportion to solve this problem. Let b represent the actual height of the bookcase in inches.

$$\frac{12 \text{ in.}}{1 \text{ in.}} = \frac{b}{2.5 \text{ in.}}$$
$$1b = 12 \times 2.5$$
$$b = 30$$

So, the actual height of the bookcase is 30 inches.

To see why the other answers are incorrect, you must remember that there are 12 inches in 1 foot and the height of the bookcase in the drawing is 2.5 inches.

Choice **a** is incorrect because a desk with a height of 12 inches would be 1 inch tall in the drawing.

Choice **b** is incorrect because an 18-inch-tall desk would be 1.5 inches tall in the drawing.

Choice **c** is incorrect because a 24-inch-tall desk would be 2 inches tall in the drawing.

112. **The correct answer is choice a.** The scale indicates that every two boxes is equal to 165 feet. From Suzette's apartment to the grocery store, there are five lengths of the scale (10 box lengths). The total distance is 165 feet × 5 = 825 feet. The length of 1 mile is 5,280 feet, so divide the total walking distance by the number of feet in 1 mile to find the distance in miles from Suzette's apartment to the grocery store: $\frac{825}{5,280} = 0.15625 \approx 0.2$ miles.
Choice **b** is incorrect because it results if 165 is multiplied by 10 instead of 5.
Choice **c** is incorrect because it results if 5,280 is divided by 825, instead of 825 divided by 5,280.
Choice **d** is incorrect because it is the distance Suzette needs to walk in feet.

113. **The correct answer is choice a.** Let x be the original price. Then the price z of the markup is given by $z = x + \frac{q}{100}x$. Solve for x:
$$z = \left(1 + \frac{q}{100}\right)x$$
$$x = \frac{1}{1 + \frac{q}{100}}z$$
$$x = \frac{100}{100 + q}z$$

Choice **b** is incorrect because you applied the $q\%$ markup to the sale price.
Choice **c** is incorrect because $q\%$ is a percent of an amount, not a quantity to be added to a given dollar amount.
Choice **d** is incorrect because the denominator is incorrect.

114. **The correct answer is choice b.** To find the distance Ralph can hike in 3 hours, first set up the ratio of the distance he can walk in a certain amount of time (45 minutes is equal to 0.75 hours): $\frac{1.3 \text{ miles}}{0.75 \text{ hours}}$. Then set up the second ratio, $\frac{d}{3 \text{ hours}}$. Set these two ratios equal to each other: $\frac{1.3}{0.75} = \frac{d}{3}$.
Choice **a** is incorrect because the left fraction compares distance to time, and the right fraction compares time to distance.
Choice **c** is incorrect because the left fraction compares shorter time to longer time, and the right fraction compares longer distance to shorter distance.
Choice **d** is incorrect because the left fraction compares shorter time to longer distance, and the right fraction compares longer distance to shorter distance.

115. The correct answer is choice b. The entire circle graph must add up to 100%. The percentages for football, baseball, basketball, and swimming add up to 70% (10 + 20 + 25 + 15 = 70). An additional 30% is needed for the entire circle graph to add up to 100%. The percentage of students who chose golf is 30%.

116. The correct answer is 3. Since 8 out of every 2,000 calculators are malfunctioning, and we are looking to see how many calculators out of 750 will be likely to have a malfunctioning problem, set up the following proportion:

$$\frac{\text{malfunctioning}}{\text{total}} = \frac{8}{2,000} = \frac{m}{750}$$

Begin to solve this proportion using cross products:

$8(750) = 2,000m$

$6,000 = 2,000m$

$m = 3$

So the company should expect to have approximately 3 returns if the stores sell all 750 calculators.

117. The correct answer is choice a. Find how many groups of two pages there are in the book: $268 \div 2 = 134$. There are 134 groups of two pages in the book. Each group will take Michael 3 minutes to read: $134 \times 3 = 402$. It will take him 402 minutes. Divide 402 by 60 to find the number of hours: $402 \div 60 = 6.7$. It will take him more than 6 hours to read the book. It is unnecessary to find the number of minutes, because there is only one answer choice of 6 hours.

118. The correct answer is choice b. The scale states that 1 inch equals 25 feet. Therefore, a tree 1.5 inches tall in a scale drawing is 25×1.5 feet tall in the real world: $25 \times 1.5 = 37.5$.

119. The correct answer is choice c. Ron sold a Jeep Cherokee at $22,995 and a Ford Escape at $22,700 for a total sale of $45,695 in one day. His daily sales goal was $42,300. Subtract the goal amount from how much he sold to get his overage: $45,695 – $42,300 = $3,395. He earns 5% commission on $3,395. So, $3,395 × 0.05 = $169.75 in commission.

Choice **a** is incorrect because it is the result of moving the decimal point of the correct answer two spaces to the left.

Choice **b** is incorrect because it is the result from moving the decimal point of the correct answer one space to the left.

Choice **d** is incorrect because it results from multiplying by 5 or 500% instead of 0.05 or 5%, thus not moving the decimal point two spaces to the left.

120. **The correct answer is 833 square feet.** Since every 10 square feet cost \$3, and the Robb family wants their maximum cost to be \$250, set up the following proportion, where f represents the number of square feet that would correspond to a cost of \$250:

$$\frac{\text{cost}}{\text{sq. ft.}} = \frac{\$3}{10} = \frac{\$250}{f}$$

Begin to solve this proportion using cross products:

$$10(250) = 3f$$
$$2,500 = 3f$$
$$\frac{2,500}{3} = 833.\overline{3} = f$$

So the Robb family can clean up to 833 square feet of carpeting without spending more than \$250.

121. **The correct answer is choice d.** If one-half of an inch represents 3 feet, then 1 inch represents twice 3 feet, or 6 feet. The dimensions of the deck on the drawing are 2 inches by 3 inches. Multiply these dimensions by 6 to arrive at the actual dimensions of 2(6), or 12, by 3(6), or 18. The actual dimensions are 12 feet by 18 feet.

122. **The correct answer is choice d.** To find the percentage increase, first find the amount of increase: \$13.00 − \$11.75 = \$1.25. Next, divide the increase amount by the original total: $\frac{1.25}{11.75} \approx 0.106$. Move the decimal point two spaces to the right to find the percentage: 10.6%. Therefore, Arnita received a 10.6% increase in pay per hour.

123. **The correct answer is choice b.** $3\frac{3}{4}$ cups = $\frac{15}{4}$ cups. Let x be the number of cups of flour needed for 90 cookies and set up the proportion $\frac{\frac{15}{4}}{36} = \frac{x}{90}$. Solving for x yields $90(\frac{15}{4}) = 36x \Rightarrow x = \frac{75}{8} = 9\frac{3}{8}$ cups.

Choice **a** is incorrect because $3\frac{3}{4} \neq \frac{9}{4}$.

Choice **c** seems to be the result of estimating. Set up a proportion to find the exact amount needed.

Choice **d** is incorrect because it is the amount needed for 72 cookies, not 90.

124. **The correct answer is choice d.** The first expression is the cost of the salad and the second term is the 6% tax on this cost; the sum is the total cost of the salad.

Choice **a** is incorrect because you did not convert 6% to 0.06 before applying it to the cost of the salad.

Choice **b** is incorrect because you forgot to include the tax.

Choice **c** is incorrect because you did not apply 6% to the cost of the salad; you don't simply add 0.06 as a term by itself without applying it to a quantity.

125. **The correct answer is choice d.** Let S be the original price. Then $x = S - \frac{p}{100}S = S(\frac{100-p}{100})$ dollars. So $S = \frac{100x}{100-p}$.

Choice **a** is incorrect because the denominator is wrong here. This would mean that the price was marked up by $p\%$, not down by $p\%$.

Choice **b** is incorrect because this is the sale price, expressed as a percentage of the original price.

Choice **c** is incorrect because you cannot cancel in fractions like this: $\frac{100x}{100-p} \neq \frac{\cancel{100}x}{\cancel{100}-p}$.

126. **The correct answer is \$45.** One way to find the cost of nine used books is to figure out the relationship between the number of books and the price of the other bundles. Do you notice a pattern? Do you see that the number of books is multiplied by 5 each time to get the price of each bundle? If we multiply 9 by 5, the answer is \$45.

Another way to find the cost of the bundle of nine books is to set up two equivalent fractions and solve for x:

$$\frac{6}{30} = \frac{9}{x}$$
$$6x = 30 \times 9$$
$$6x = 270$$
$$x = \frac{270}{6}$$
$$x = 45$$

127. **The correct answer is choice d.** First, convert 6 miles per hour to feet per minute:

$$\frac{6 \text{ mi.}}{1 \text{ hour}} \times \frac{5{,}280 \text{ ft.}}{1 \text{ mi.}} \times \frac{1 \text{ hour}}{60 \text{ min.}} = 528 \text{ feet per minute}$$

Let x be the number of minutes it takes to jog 5,000 feet. Solve the equation $528x = 5{,}000$ to get $x = \frac{5{,}000}{528} \approx 9.5$ minutes.

Choice **a** is incorrect because you divided 5,000 by 6. You must first convert units appropriately.

Choice **b** is incorrect because you mistakenly used 1,760 feet instead of 5,280 in the mile-to-feet conversion, and you looked for the number of seconds, not the number of minutes.

Choice **c** is incorrect because you mistakenly used 1,760 feet instead of 5,280 in the mile-to-feet conversion.

128. **The correct answer is choice a.** First, apply the 15% coupon to get $75.50 − $75.50(0.15) = $64.18. Now apply the 6% tax to this amount to get $64.18 + $64.18(0.06) = $68.03. Now apply the 20% tip to this amount to get the final bill: $68.03 + $68.03(0.20) = $81.64.

Choice **b** is incorrect because you did not apply the tax.

Choice **c** is incorrect because you did not include tax or tip in this amount.

Choice **d** is incorrect because you did not include the tip in this amount.

129. **The correct answer is choice d.** 46% of 327,577,529 is approximately 150,685,663.3. So rounding to the hundred thousands place gives 150,700,000.

Choice **a** is incorrect because you rounded to the 10 millions place.

Choice **b** is incorrect because you rounded to the 10 thousands place.

Choice **c** is incorrect because you rounded (incorrectly) to the 10 thousands place.

130. **The correct answer is choice d.** Let x be the desired mass. Set up a proportion and solve for x:

$$\frac{5 \text{ containers}}{3.2 \text{ kg}} = \frac{8 \text{ containers}}{x \text{ kg}}$$
$$5x = 8(3.2) = 25.6$$
$$x = 5.12 \text{ kg} = 5{,}120 \text{ g}$$

Choice **a** is incorrect because you set up the proportion incorrectly by inverting one of the fractions.

Choice **b** is incorrect because you used the wrong units; this should be 5.12 kg, not 5.12 g.

Choice **c** is incorrect because this would correspond to ten containers, not eight.

4

Variables, Linear Equations, and Linear Inequalities

Now that you've tested your mastery of the important foundational work in the past three chapters, you are ready to move on to algebra. Algebra is an organized system of rules that are used to help solve problems for *unknowns*. It is because of algebra that engineers can build bridges, cell phones, and engines. In all of these cases, equations are written to determine the values of unknowns. In order to correctly solve the more advanced algebraic questions presented in the next chapter, you will need to be comfortable and confident answering questions on the following topics presented in this chapter:

- Adding, subtracting, multiplying, and dividing algebraic expressions

- The distributive property and linear expressions

- Evaluating algebraic expressions through substitution

- Solving linear equations

- Modeling word problems with algebra

131. Which algebraic expression represents each description?
(Drag-and-drop question)

7 less than twice a number y: _____

Twice the sum of 7 and a number y: _____

$7 - 2y$

$(7 + y)^2$

$2y + 7$

$2 \times 7 + y$

$2(7y)$

$7 + 2y$

$2y - 7$

$(7 + y)^2$

$2(7 - y)$

132. If n is any negative integer, complete the following statements:
(*Note:* This would be a drop-down menu question on the GED test—to complete this practice question choose from *sometimes/always/never*.)

A. The expression $\frac{(-n)^4}{n}$ will *sometimes/always/never* be positive.

B. The expression $\frac{-(n)^4}{n}$ will *sometimes/always/never* be positive.

133. What is the value of $4x^2 + 3(1 - x)$, when $x = -3$?
   a. 48
   b. 156
   c. −12
   d. 30

134. Simplify the expression $5x + 3(x - 4)^2$. _____

135. Expand and simplify the following expression:
$7(x + 2y - 3) - 3(2x - 4y + 1)$
   a. $x + 2y - 18$
   b. $13x - 2y - 2$
   c. $x + 26y - 24$
   d. $x + 6y - 4$

136. If the sum of two polynomials is $8p^2 + 4p + 1$ and one of the polynomials is $8p^2 - 2p + 6$, what is the other polynomial?
  a. $6p - 5$
  b. $2p + 7$
  c. $16p^2 - 2p + 7$
  d. $16p^2 + 2p + 7$

137. Which of the following is equivalent to $2x(3xy + y)$?
  a. $6x^2y + 2xy$
  b. $6xy + 2xy$
  c. $5x^2y + 2x + y$
  d. $3xy + 2x + y$

138. Which expression is the equivalent of $32x^2 + 4x - 8$?
  a. $32(x^2 + 4x - 8)$
  b. $4x(x + x - 8)$
  c. $4(x^2 + x - 2)$
  d. $4(8x^2 + x - 2)$

139. Which of the following is a factored form of $10x^4y^6 - 5x^3y$?
  a. $5xy^5$
  b. $5x^3y$
  c. $5x^3y(2xy^5 - y)$
  d. $5x^3y(2xy^5 - 1)$

140. A host for a party decides to buy three balloons for every guest, plus 20 balloons to decorate the hall. If g represents the number of guests invited to the party and b represents the total number of balloons to be purchased, which equation shows the relationship between the number of balloons (b) and the number of guests (g)?
  a. $b = 3(g + 20)$
  b. $b = 3g + 20$
  c. $b = 60g$
  d. $b = 23g$

141. Farhiyo and Jen sold T-shirts for a campus club last Saturday. The club made $550 from selling these T-shirts. After donating some of the money to a local shelter, the club made $100 more than it donated. How much money did the club donate? _____

142. Aaron owns a pretzel stand. After observing sales patterns for a few months, he realizes that he needs to have three times as much cheese as he does ranch dressing to fulfill customers' orders. For every 48 ounces of cheese Aaron buys, how much ranch dressing should he buy?

   a. 144 oz.
   b. 24 oz.
   c. 12 oz.
   d. 16 oz.

143. Johanna and Paolo just finished a three-day promotional event for their new business. They distributed flyers to businesses and homes in the neighborhood to let the public know about their new shop. They printed a total of 1,000 flyers for this promotional effort and distributed them over the course of three days. They have x flyers left over. If they are expecting a 15% response rate—meaning that, of the flyers handed out, 15% will bring in one person—which expression illustrates how many more customers they are expecting in the near future as a result of this promotional effort?

   a. $15(1,000 - x)$
   b. $15(1,000x)$
   c. $15 + 1,000 - x$
   d. $0.15(1,000 - x)$

144. The product of 16 and one-half of a number is 136. Find the number. _____

145. Enya makes $1,500 a month after taxes. She has $1,000 of expenses, including rent, utilities, and food. She wants to save at least $325 each month so that she can buy a used car in a year or two. Which inequality accurately represents how much spending money, m, Enya can potentially have each month while still achieving her saving goal?

   a. $m \geq \$175$
   b. $m \geq \$325$
   c. $m \leq \$325$
   d. $m \leq \$175$

146. Find the solution set to the compound inequality: $3 < 4x - 9 < 23$
   a. $x < 8$
   b. $x > 8$ or $x < 3$
   c. $3 > x > 8$
   d. $3 < x < 8$

147. The product of two consecutive integers is 42. If the smaller integer is x, which of the following equations must be true?
   a. $x + 1 = 42$
   b. $x^2 + x = 42$
   c. $2x + 1 = 42$
   d. $2x^2 + x = 42$

148. A real-estate agent has found that the asking price of a home in his area can be estimated by taking the square footage, multiplying by 84, and adding 1,065. If the square footage is represented by S and the asking price by P, then which of the following formulas represents this estimation?
   a. $P = 1,149S$
   b. $P = 84(S + 1,065)$
   c. $P = 84S + 1,065$
   d. $P = S + 1,149$

149. Suppose that for a rational number x, $3(x - 5) = 3$. Select which of the following must be true.
   a. $x - 5 = 1$
   b. $3x - 15 = 9$
   c. $x = 5$
   d. $3x = 8$

150. Which of the following represents the solution set of the inequality $4x - 9 < 3x + 1$?
   a. $x < -\frac{8}{7}$
   b. $x < -8$
   c. $x < 10$
   d. $x < \frac{10}{7}$

151. $\frac{2}{x(x-1)} + \frac{1}{x-1} =$

    a. $\frac{3}{2x(x-1)}$

    b. $\frac{2+x}{x(x-1)}$

    c. $\frac{3}{x(x-1)}$

    d. $\frac{2}{x-1}$

152. Susie wants to find the value of $y = 7x - 2$ when $x = 5$. She solves the equation and finds y to be equal to 25. What did Susie do wrong?

    a. She substituted 5 for x and multiplied by 7 before subtracting 2.

    b. She substituted 4 for x and multiplied by 7 before subtracting 2.

    c. She subtracted 2 from 7 before substituting 5 for x and multiplying.

    d. She subtracted 2 from x before substituting and multiplying.

153. If $\frac{3}{4}x = 12$, then $x =$

    a. 9

    b. $11\frac{1}{4}$

    c. $12\frac{3}{4}$

    d. 16

154. Solve for x: $3x - 8 = 8x - 20$

    a. $x = -2.4$

    b. $x = 2$

    c. $x = 2.4$

    d. $x = 5.6$

155. Joseph owns v video games. Harry owns 10 fewer than 2 times the number of video games that Joseph owns. Which expression represents the number of video games that Harry owns in terms of v?

    a. $10v - 2$

    b. $2v - 10$

    c. $2(v - 10)$

    d. $10(v - 2)$

156. As part of a game, Gilbert must take a number and use a special procedure to come up with a new number. To come up with his new number, Gilbert takes the original number, cubes it, adds 5 to it, and finally multiplies it by 2. If the original number is represented by x, which of the following represents Gilbert's new number?
   a. $2(3x + 5)$
   b. $2(x^3 + 5)$
   c. $2x^3 + 5$
   d. $x^6 + 5$

157. The sum of a number n and 4 is less than 5 times the number m. If m is 6, which of the following is true?
   a. n is greater than 6
   b. $n + 4$ is less than 26
   c. n is less than 26
   d. n is equal to 26

158. Which inequality represents the statement "Five times a number is less than 45"?
   a. $5x < 45$
   b. $45 - 5x$
   c. $5x > 45$
   d. $5x - 45$

159. Clare rented a car while on her vacation. The ABC car rental company charged her $45 plus $0.50 per mile. Which equation can be used to correctly calculate the cost of renting the car? Assume that C is the cost of renting the car and m is the number of miles driven.
   a. $C = 0.50m + 45$
   b. $C = 0.50(m + 45)$
   c. $C = m(0.50 + 45)$
   d. $C = 45m + 0.50$

160. Solve the following equation for x:
   $$-3x + 6 = 3$$
   a. $x = -3$
   b. $x = -1$
   c. $x = 0$
   d. $x = 1$

161. Find the sum of $6x + 5y$ and $3x - 5y$.
   a. $9x + 10y$
   b. $9x - 10y$
   c. $10y$
   d. $9x$

162. Three apples and twice as many oranges add up to one-half the number of cherries in a fruit basket. How many cherries are there?
   a. 11
   b. 18
   c. 21
   d. 24

163. A factory is able to produce at least 16 items, but no more than 20 items, for every hour the factory is open. If the factory is open for 8 hours a day, which of the following are possibly the numbers of items produced by the factory over a 7-day work period?

Select all of the correct possibilities from the list and write them in the box.

128
150
850
910
1,115

```
┌─────────────────────────────┐
│                             │
└─────────────────────────────┘
```

164. Which of the following represents the solution set of the inequality $x + 2 > 5$?
   a. $\{x: x > 10\}$
   b. $\{x: x > 7\}$
   c. $\{x: x > 3\}$
   d. $\{x: x > 2.5\}$

165. What value of x satisfies the system of equations $x - 2y = 8$ and $x + 2y = 14$?
    a. -6
    b. 11
    c. There are infinitely many values of x that satisfy this system.
    d. There are no values of x that satisfy this system.

166. $(x^2 + 5) - (x^2 - x) =$
    a. $5 + x$
    b. $5 - x$
    c. $2x^2 - 5x$
    d. $2x^2 + x + 5$

167. Edward's school is having a raffle contest. People can buy a red ticket for $10 for a chance to win a new TV or a blue ticket for $5 for a chance to win a brand-new bicycle. Edward sold a record high of 130 tickets for a total of $1,100. How many of each color ticket did Edward sell? _____

168. Solve for x: $4x - 7 = 37$
    a. $x = 7.5$
    b. $x = 10$
    c. $x = 11$
    d. $x = 40$

169. Robert turned 4 today and Garrett turned 2 today. Which equation represents the relationship between their ages at any time in their lives? Use R to represent Robert's age and G to represent Garrett's age.
    a. $2G = R$
    b. $2R = G$
    c. $G = R + 2$
    d. $G = R - 2$

170. If $2y + 11 = 25$, what is the value of y?
    a. 18
    b. $\frac{11}{23}$
    c. 6
    d. 7

171. Solve the equation for z:

$$3z - 2 = 7 + 4z$$

a. $z = -9$

b. $z = -5$

c. $z = 5$

d. $z = 9$

172. Suppose that a coat costs $25 more than twice the cost of a sweater. If c represents the cost of the coat, and s represents the cost of the sweater, which equation can you use to find the cost of the coat?

a. $s = c + 25$

b. $c = s + 25$

c. $s = 2c + 25$

d. $c = 2s + 25$

173. Multiply these binomials:

$$(7j + 2k)(3j - k)$$

a. $21j^2 - 2k^2$

b. $21j^2 + 13jk - 2k^2$

c. $21j^2 - jk - 2k^2$

d. $21j^2 - 13jk - 2k^2$

174. Add the polynomials:

$$(2t^3 - 7t^2 + 5) + (4t^2 + 11t - 9)$$

a. $6t^3 + 4t^2 - 4$

b. $6t^3 - 4t^2 + 4$

c. $2t^3 - 3t^2 + 11t + 4$

d. $2t^3 - 3t^2 + 11t - 4$

175. Jose and Lavina supervise volunteer firefighters. They need to buy new equipment such as fire helmets and have raised $2,700 from the community to do this. If each helmet costs $300, which expression shows how many helmets can be purchased?

a. $\frac{300}{2,700} = n$

b. $300n = 2,700$

c. $\frac{n}{2,700} = 300$

d. $300 = 2,700n$

176. Dave and 12 others threw a surprise party for their friend Carolyn at the group's favorite restaurant for a total of $305 plus room rental fee. Dave offered to cover the $45 fee to rent the lower-level room in addition to helping the others pay for appetizers. If everyone contributed the same amount toward appetizers, which equation could be used to figure out how much each person paid?

 a. $305 = 13x - 45$
 b. $305 = 12x + 45$
 c. $305 = 13x + 45$
 d. $305 = 12x - 45$

177. Suewand makes her own trail mix. She buys almonds, raisins, and chocolate pieces in bulk once a month. If she buys 3 lbs. of almonds at $8.99 per pound and 1.5 lbs. of raisins at $2.50 per pound, which expression illustrates how many pounds of chocolate pieces Suewand can buy at $3.99 per pound if she has only $37.00 to spend total? *Note:* There is no tax on food where Suewand lives.

 a. $30.72 + 3.99x \le 37$
 b. $30.72 - 3.99x \le 37$
 c. $30.72 - 3.99x \ge 37$
 d. $30.72 + 3.99x \ge 37$

178. For which of the values of a does the following system have no solution?

$$\begin{cases} y = ax - 3 \\ ay = x + 1 \end{cases}$$

 a. $a = \frac{1}{2}$
 b. $a = -1$
 c. $a = -2$
 d. $a = 4$

179. Will invests a portion of $6,500 in a money market CD that earns 3% interest annually, and deposits the remainder in a savings account that earns 1.4% interest annually. He earns $100.50 in interest in the first year. Let x represent the amount Will invests in the 3% CD. Which of the following equations could be used to determine the amount he invested in the CD and the amount he deposited in the savings account?

a. $0.03x + 0.014(6,500 - x) = 100.50$
b. $0.03x + 0.014(6,500 - x) = 6,500(0.044)$
c. $3x + 1.4(6,500 - x) = 100.50$
d. $1.4x + 3(x - 6,500) = 10,050$

180. Sofia has a collection of nickels, dimes, and quarters. She has two fewer nickels than quarters, and the number of dimes is one more than twice the number of quarters. If her total amount of money is $3.50, how many nickels are in her collection?

a. 5
b. 7
c. 14
d. 15

181. Which of the following expressions is equivalent to $[(1 - 3x^2)(1 + 3x^2)]^2$?

a. $81x^8 - 18x^4 + 81x^8$
b. $1 - 81x^8$
c. $81x^6 - 18x^4 + 1$
d. $1 + 36x^8$

182. What is the solution set for the inequality $-x - 3[4(1 - 2x) + 5] > x$?

a. $\{x \mid x > \frac{8}{9}\}$
b. $\{x \mid x > \frac{27}{22}\}$
c. $\{x \mid x > -\frac{7}{10}\}$
d. $\{x \mid x > \frac{22}{27}\}$

183. Solve for y: $\frac{7}{x} - \frac{3}{y} = \frac{4}{2x}$

    a. $y = \frac{3}{5}x$

    b. $y = -\frac{15}{14}x$

    c. $y = \frac{5}{3}x$

    d. $y = -x$

184. John buys turkey sandwiches and roast beef sandwiches at a local sandwich shop. A turkey sandwich costs \$5.50, and a roast beef sandwich costs \$6.75. If he buys 12 sandwiches all told and spends \$72.25, which system of equations could be used to determine the number of each type of sandwich he purchased?

    a. $\begin{cases} x + y = 72.25 \\ 5.50x + 6.75y = 12 \end{cases}$

    b. $\begin{cases} 12.25(xy) = 72.25 \\ x + y = 12 \end{cases}$

    c. $\begin{cases} x + y = 12 \\ 5.50x + 6.75y = 72.25 \end{cases}$

    d. $\begin{cases} 550x + 675y = 72.25 \\ x + y = 12 \end{cases}$

185. What is the solution of the system $\begin{cases} 2x - 1 = y \\ -2y + 1 = x \end{cases}$?

    a. $x = 1, y = 0$

    b. $x = \frac{3}{5}, y = \frac{1}{5}$

    c. $x = \frac{2}{5}, y = -\frac{1}{5}$

    d. $x = 1, y = 1$

186. Which of the following expressions is equivalent to $(2x^2 + 3)(2 - 4x^2)$?

    a. $-8x^2$

    b. $8x^4 + 6$

    c. $-8x^4 - 8x^2 + 6$

    d. $8x^4 + 8x^2 - 6$

187. Solve for x: $-2(x + 2) - 1 = 2(2 - x)$.

    a. -3

    b. $\frac{1}{4}$

    c. 9

    d. no solution

188. The fuel efficiency of a car is measured using the equation $F = \frac{M}{g}$, where M is the number of miles and g is the number of gallons used to travel that number of miles. If Sam's fuel efficiency is 56.1 miles per gallon and he drove 151.47 miles, how many gallons of fuel did he use?

    a. 95.4

    b. 0.37

    c. 27.0

    d. 2.70

189. Which of the following expressions is NOT equivalent to $x(2x - 3) - 3(4 - 2x^2) + 6x^2$?

    a. $6x^2 + x(2x - 3) - 3(4 - 2x^2)$

    b. $-[3(4 - 2x^2) - 6x^2 - x(2x - 3)]$

    c. $14x^2 - 3(x + 4)$

    d. $2x^2 - 3x - 12$

190. A merchant at the farmers' market sells plums and peaches. If the cost for 5 plums and 6 peaches is $6.75 and the cost of 8 plums and 3 peaches is $7.50, which system can be used to determine the price of 1 plum and 1 peach?

    a. $\begin{cases} x + y = 11 \\ 8x + 3y = 14.25 \end{cases}$

    b. $\begin{cases} 5x + 6y = 7.50 \\ 8x + 3y = 6.75 \end{cases}$

    c. $\begin{cases} 5x + 3y = 6.75 \\ 8x + 6y = 7.50 \end{cases}$

    d. $\begin{cases} 5x + 6y = 6.75 \\ 8x + 3y = 7.50 \end{cases}$

191. Which of the following systems has no solution?

a. $\begin{cases} 4x = 2y - 1 \\ 6x = 3y + 4 \end{cases}$

b. $\begin{cases} 2x = 7y - 3 \\ 4x = 14y - 6 \end{cases}$

c. $\begin{cases} 2x - 7 = 1 \\ 3y + 5 = 2 \end{cases}$

d. $\begin{cases} 5y - x = 0 \\ 10y = 2x \end{cases}$

192. For which of the following inequalities is the solution set all real numbers?

a. $x^2 - 4 \geq 0$
b. $x(x + 5) < 0$
c. $x^2 > -9$
d. $(x - 3)^2 < 0$

193. Assume a is a non-zero real number. Solve the following system:

$$\begin{cases} 2x + y = a \\ x - y = 2a \end{cases}$$

a. $x = 0, y = a$
b. $x = -a, y = a$
c. $x = a, y = -a$
d. $x = a, y = 3a$

194. Micah's tennis record suggests that whenever he serves the ball, the probability that he will score the point is 0.35, the probability that he will have one fault is 0.47, and the probability that he will do neither is 0.61. What is the probability that Micah either scores the point or has one fault?

a. 0.82
b. 0.66
c. 0.35
d. 0.49

195. Which of the following systems has infinitely many solutions?

a. $\begin{cases} y = 3x + 1 \\ y = 3x - 1 \end{cases}$

b. $\begin{cases} x = -4y \\ y = 4x \end{cases}$

c. $\begin{cases} 2x - y = 5 \\ -x + y = -1 \end{cases}$

d. $\begin{cases} 2x - 3y = -1 \\ -4x + 6y = 2 \end{cases}$

196. Solve for x: $4(3 - 2x) - 2(1 - 3x) = -2$.

a. $\frac{12}{5}$

b. $\frac{1}{6}$

c. -4

d. 6

197. The volume V of a right circular cone with base radius r and height h is given by $V = \frac{1}{3}\pi r^2 h$. What is the formula for the radius r in terms of V and h?

a. $r = \sqrt{3V\pi h}$

b. $r = 6\pi h V$

c. $r = \frac{6V}{\pi h}$

d. $r = \sqrt{\frac{3V}{\pi h}}$

198. A shopper buys a mixture of CDs and vinyl albums online. The cost per CD is $11.50 and the cost per vinyl album is $15.75. She buys three fewer CDs than two times the number of vinyl albums and spends $275.50 on the entire purchase before tax. Which equation can be used to determine the number of CDs and number of vinyl albums that she purchases?

a. $11.50x + 15.75y = 275.50$

b. $1{,}150(2x + 3) + 1{,}575x = 27{,}550$

c. $15.75x + 11.50(2x - 3) = 275.50$

d. $27.25[x + (2x + 3)] = 275.50$

199. $-x^2(x + 1) - (x^3 + 4x^2) =$

    a. $-6x^3 - x^2$

    b. $-2x^3 - 5x^2$

    c. $-2x^3 + 3x^2$

    d. $-2x^3 + 4x^2 + 1$

Answers and Explantions

131. **The correct answers are $2y - 7$ and $(7 + y)^2$.** *Less than* means subtraction in the reverse order from that stated and *twice a number y* means $2y$, so the first expression, *7 less than twice a number y*, is represented by $2y - 7$. In the second expression, the phrase *sum of 7 and a number y* indicates that $7 + y$ should be put inside a set of parentheses before being multiplied by 2. Therefore, the expression that represents *twice the sum of 7 and a number y* is represented by the expression $(7 + y)^2$. The expression $2(7 + y)$ would also have worked, but this option was not on the list.

132. **The correct answers are never and always.** Think about the rules for how negative numbers are influenced by exponents and parentheses. For statement A, $\frac{(-n)^4}{n}$, the numerator will always be positive since the negative n will get canceled out by the negative sign that is within the parentheses. The denominator will always be negative. The quotient of a positive divided by a negative is never positive. For statement B, $\frac{-(n)^4}{n}$, the numerator will always be negative since n^4 will also yield a positive value and then the negative sign on the outside of the parentheses will make the numerator negative. The denominator will always be negative. The quotient of a negative divided by a negative is **always** positive.

133. **The correct answer is choice a.** We cannot multiply 4×-3 and then square the answer because exponents come before multiplication in the order of operations. Plus, there are parentheses in the second half of the problem, and PEMDAS indicates we do that first.

$$4(-3)^2 + 3(1 - -3) = 4(-3)(-3) + 3(1 - -3) = 4 \times 9 + 3(1 + 3)$$
$$= 36 + 3(4) = 36 + 12 = 48$$

Choice **b** is not the correct answer because 156 is the final answer received when 4 is multiplied to the -3 and *then* squared, rather than doing the squaring first. Choice **c** is not the correct answer because -12 is the answer when you mistake $(-3)^2$ for -6 instead of 9. Choice **d** is not the correct answer because 30 is the result of mistaking $(1 - -3)$ as -2 instead of 4.

134. **The correct answer is $3x^2 - 19x + 48$.** To simplify $5x + 3(x - 4)^2$ we must first think about what $(x - 4)^2$ means:

$$5x + 3(x - 4)^2 = 5x + 3(x - 4)(x - 4)$$

Now use FOIL to expand $(x - 4)(x - 4)$ to be $(x^2 - 8x + 16)$ and plug that into the equation:

$$5x + 3(x^2 - 8x + 16)$$

Now distribute the 3 to all the terms inside the parentheses using multiplication:

$$5x + 3x^2 - 24x + 48$$

Last, combine the two x terms.

$$3x^2 - 19x + 48$$

135. **The correct answer is choice c.** This problem requires us to use the distributive property for two parts of the expression. First, we need to distribute 7 to each of the factors in the quantity after it: x, $2y$, and -3. This gives us $7x + 14y - 21$. Next, we need to distribute the -3 to each of the terms in the quantity after it: $2x$, $-4y$, and 1. This gives us $-6x + 12y - 3$. Now, we need to combine like terms to simplify the expression: $7x + 14y - 21 - 6x + 12y - 3 = x + 26y - 24$. Choice **a** did not distribute the negative with the 3 in the second half of the problem. Choice **b** reflects incorrectly distributing the coefficients and negatives to the other terms. Choice **d** did not distribute the coefficient to each term.

136. **The correct answer is choice a.** Deciding whether to add or subtract is the trick to answering this question.

Sum = first polynomial + second polynomial

Since the sum is already given, do not add. Instead, subtract the polynomial given from the sum to find the other polynomial. Be sure to combine like terms, distribute the subtraction, and be careful with negatives.

$$(8p^2 + 4p + 1) - (8p^2 - 2p + 6) = (8p^2 - 8p^2) + (4p - (-2p)) + (1 - 6)$$
$$= 0 + (4p + 2p) + (-5) = 6p - 5$$

Choice **b** is incorrect because you simply added the first-degree terms and the constant terms of the two given polynomials, rather than seeing that you had to subtract one polynomial from the other.

Choice **c** is incorrect because you attempted to find the sum of the two given polynomials (but made a sign mistake with your $-2p$), but this is not what the question was asking.

Choice **d** is incorrect because you attempted to find the sum of the two given polynomials, rather than finding the missing polynomial needed to make the given sum.

137. **The correct answer is choice a.** $2x(3xy + y) = 2x(3xy) + 2x(y) = 6x^2y + 2xy$.

Choice **b** is incorrect because you forgot to square x when multiplying $2x$ by $3xy$.

Choice **c** is incorrect because you didn't correctly multiply 2 and 3, which is 6, and you added $2x$ and y instead of multiplying them.

Choice **d** is incorrect because you added all the terms instead of distributing the $2x$ to both terms within the parentheses.

138. **The correct answer is choice d.** This answer is equivalent to the original because each of the terms was divided by 4, and then accurately written as a product of 4 and the quantity $8x^2 + x - 2$. This, in effect, is the un-distribution of a 4 from each term. Choices **a** and **b** incorrectly factor out 32 and $4x$, respectively, which leaves the remaining terms incorrect. Choice **c** looks very similar to choice **d**, but the 4 is not correctly factored out of the term $32x^2$.

139. **The correct answer is choice d.** To factor means to divide out the largest factor that all of the terms have in common.

First, consider the coefficients: 5 is a factor of both coefficients, so divide the 5 out.

Next, consider the variable x: x^3 is a factor of both terms, so divide out the x^3.

Finally, consider the variable y: The second term has y raised to only the first power, so y is the largest factor.

After determining the largest factor, rewrite the expression by dividing each term by the factor:

$5x^3y(2xy^5 - 1)$

Choice **a** is incorrect because $5xy^5$ is not even a common factor for the expression $10x^4y^6 - 5x^3y$, since both terms do not have y^5. (Plus this is a single monomial and not a factored binomial.)

Choice **b** is incorrect because, although $5x^3y$ is the greatest common factor for the expression $10x^4y^6 - 5x^3y$, it is a single monomial and you forgot to use it to create a factored binomial that is equivalent to the original expression.

Choice **c** is incorrect because $5x^3y$ times y would yield $5x^3y^2$ for the second term in the binomial, which is not equivalent to the original expression.

140. **The correct answer is choice b.** The party host must buy three balloons for each guest, so $3g$ represents the correct number of balloons for all the guests. The host must also buy 20 more balloons for the hall, so the total number of balloons is the number needed for the guests plus 20 more. This is the equation $b = 3g + 20$.

Choice **a** is incorrect because the expression $3(g + 20)$ will end up being $3g + 60$ balloons, which will be 3 balloons for every guest, plus an additional 60 balloons, rather than an additional 20 balloons.

Choice **c** is incorrect because $60g$ represents that 60 balloons would be purchased for every guest.

Choice **d** is incorrect because $23g$ represents that 23 balloons would be purchased for every guest.

141. **The correct answer is $225.** A system of equations is needed to solve this problem. If C = the amount of money the club profits and D = the amount of money donated, the following two equations are true:

$$C + D = 550$$
$$C - 100 = D$$

When $C - 100$ is substituted into the first equation for D, the equation reads:

$$C + C - 100 = 550$$
$$2C - 100 = 550$$
$$2C = 650$$
$$C = 325$$

When this value is substituted for C in the second equation, the equation reads $325 - 100 = 225$.

142. **The correct answer is choice d.** Since we know that Aaron needs three times as much cheese as ranch dressing, we know that the 48 ounces of cheese are three times the needed ranch dressing. Either multiply 48 by $\frac{1}{3}$ or divide 48 by 3. They are essentially doing the same thing. The answer is 16 ounces of ranch.

Choice **a** is incorrect because you calculated that Aaron needs three times more ranch dressing than cheese, when the question states the opposite: that he needs three times more cheese than ranch dressing.

Choice **b** is incorrect because 24 ounces of ranch dressing would give Aaron twice as much cheese as ranch dressing; however, he needs three times as much cheese.

Choice **c** is incorrect because 12 ounces of ranch dressing would give Aaron four times as much cheese.

143. **The correct answer is choice d.** In order to find out how many flyers Johanna and Paolo passed out, we need to subtract x, the amount left over, from the total, 1,000. Once we get that value, we multiply by 15% by changing 15% to a decimal, 0.15. Multiply 0.15 by $(1,000 - x)$ to get the number of new customers they will anticipate.

Choice **a** is incorrect because you set the expression up correctly, but you forgot to write 15% as a decimal and instead represented it as a whole number.

Choice **b** is incorrect because it makes no sense to multiply the number of flyers they started with by the number of flyers they did not distribute, x. You also forgot to write 15% as a decimal instead of a whole number.

Choice **c** is incorrect because you forgot to group $1,000 - x$ within a set of parentheses and you also forgot to write 15% as a decimal instead of a whole number.

144. **The correct answer is $x = 17$.** Let x equal the number sought. The word *product* tells us to multiply 16 by one-half x, or $(16)(0.5x)$, which we set equal to 136. Therefore, $(16)(0.5x) = 136$, which reduces to $8x = 136$, resulting in $x = 17$.

145. **The correct answer is choice d.** This scenario indicates that Enya wants to save *at least* $325 per month. That means that after starting with $1,500 of income, and subtracting $1,000 for expenses and also subtracting m dollars for spending money, she wants to have at least $325 left over. This can be represented as follows:

$1,500 - $1,000 - m \geq 325, where m = spending money

Simplifying the left side yields $500 - m \geq 325

Now move the m over to the right side and bring the $325 over to the left side for a final inequality of $175 \geq m$, which can also be written as $m \leq 175. This means that Enya's monthly spending money must be $175 or less in order to meet her long-term goal of purchasing a car.

Answer choice **a** doesn't make sense because if she spends more than $175 per month, her savings will go below her $325/month goal. This would be the answer for a student who divided by –1 to get m alone and then forgot to change the direction of the inequality sign. Answer choices **b** and **c** don't work because here the "saving money" is being used as the "spending money."

146. **The correct answer is choice d.** The goal is to isolate x in the middle of the two inequalities. What you do to the middle of the inequality you must do to the left and right sides as well. The first step is to add 9 to all three parts of the compound inequality. Then divide all three parts by 4 to get x fully alone:

$$3 < 4x - 9 < 23$$
$$\underline{+9 \qquad +9 \quad +9}$$
$$12 < 4x \qquad < 32$$
$$\frac{12}{4} < \frac{4x}{4} < \frac{32}{4}$$
$$3 < x < 8$$

x is the set of all numbers greater than 3 and less than 8.

Answer choices **b** and **c** have aspects of the correct solution, but mistakes have been made with the directions of their inequality symbols.

Choice **a** has only half of the solution, $x < 8$, but this must be paired with $x > 3$ as well.

147. **The correct answer is choice b.** If the first integer is x, then the second integer is $x + 1$ and their product is $x(x + 1) = x^2 + x = 42$.

Choice **a** is incorrect. The second integer would be $x + 1$, but the product of both integers should be included in the equation.

Choice **c** is incorrect. The question asks for the *product*, not the *sum*.

Choice **d** is incorrect. While there are two integers, neither integer will be represented by $2x$.

148. **The correct answer is choice c.** Multiplying by 84 is the first step, and this is represented by $84S$. The 1,065 is added to this term, leading to the model $P = 84S + 1,065$.

Choice **a** is incorrect. This model represents multiplying the square footage by 1,149.

Choice **b** is incorrect. This model represents multiplying by 84 as the last step and would produce different results.

Choice **d** is incorrect. This model represents just adding 84 and then 1,065 to the square footage.

149. **The correct answer is choice a.** Dividing both sides by 3 shows that $x - 5 = 1$.

Choice **b** is incorrect. Multiplying the expressions on the left side yields $3x - 15 = 3$, *not* $3x - 15 = 9$.

Choice **c** is incorrect. Dividing both sides by 3 shows that $x - 5 = 1$. This can be further reduced to $x = 6$, *not* $x = 5$.

Choice **d** is incorrect. Multiplying the expressions on the left side yields $3x - 15 = 3$. This can be further reduced to $3x = 18$, *not* $3x = 8$.

150. **The correct answer is choice c.** After subtracting $3x$ from both sides, the resulting inequality is $x - 9 < 1$. Adding 9 to both sides results in the final solution of $x < 10$.

Choice **a** is incorrect. Since the sign of $3x$ is positive, it should be subtracted from both sides. Similarly, in the next step, the 9 should be added to both sides since it is subtracted from $4x$.

Choice **b** is incorrect. After subtracting the $3x$ from both sides, the 9 should be added to both sides since it is subtracted from $4x$.

Choice **d** is incorrect. Since the sign of $3x$ is positive, it should be subtracted from both sides.

151. **The correct answer is choice b.** $\frac{2}{x(x-1)} + \frac{1}{x-1} = \frac{2}{x(x-1)} + \frac{x}{x(x-1)} = \frac{2+x}{x(x-1)}$.

Choice **a** is incorrect. The fractions must have a common denominator before they can be added, and once they do, only the numerators are combined.

Choice **c** is incorrect. While the common denominator is $x(x-1)$, $\frac{1}{x-1} \neq \frac{x}{x(x-1)}$.

Choice **d** is incorrect. The x terms in the numerator and denominator after finding a common factor are not factors and therefore cannot be canceled.

152. **The correct answer is choice c.** When solving an equation, first replace any variables with their values, and then follow the order of operations—always perform multiplication and division before addition and subtraction. Susie subtracted before multiplying. She subtracted 2 from 7 and arrived at 5, and then she multiplied 5 by the value of x, 5, to arrive at 25.

Choice **a** is incorrect because if Susie had followed the steps described in this choice, she would have arrived at the correct answer, 33, not 25.

Choice **b** is incorrect because if Susie had followed the steps described in this choice, she would have arrived at 26, not 25.

Choice **d** is incorrect because if Susie had followed the steps described in this choice, she would have arrived at 21, not 25.

153. **The correct answer is choice d.** To isolate the x, multiply both sides of the equation by the reciprocal of $\frac{3}{4}$. Thus $x = \frac{4}{3}(12) = 16$.

Choice **a** is incorrect. To cancel out the $\frac{3}{4}$, both sides should be multiplied by the reciprocal instead of the original fraction.

Choice **b** is incorrect. Subtracting the fraction from both sides will not isolate the x since the x is multiplied by the fraction.

Choice **c** is incorrect. Adding the fraction to both sides will not isolate the x since the x is multiplied by the fraction.

154. **The correct answer is choice c.** You can use the following steps to solve this equation:

$$3x - 8 = 8x - 20$$
$$3x - 8 + 8 = 8x - 20 + 8$$
$$3x = 8x - 12$$
$$3x - 8x = 8x - 12 - 8x$$
$$-5x = -12$$
$$\frac{-5}{-5}x = \frac{-12}{-5}$$
$$x = 2.4$$

Choice **a** is incorrect because if x were –2.4, then $3x - 8$ would be –15.2, and $8x - 20$ would be –0.8.

Choice **b** is incorrect because if x were 2, then $3x - 8$ would be –2, and $8x - 20$ would be –4.

Choice **d** is incorrect because if x were 5.6, then $3x - 8$ would be 8.8, and $8x - 20$ would be 24.8.

155. **The correct answer is choice b.** The phrase *10 fewer than* implies that 10 should be subtracted from the next stated term. That term is *2 times the number of video games that Joseph owns*, or $2v$.

Choice **a** is incorrect. This expression represents 2 fewer than 10 times the number of video games Joseph owns.

Choice **c** is incorrect. This expression represents 2 times 10 fewer than the number of video games Joseph owns.

Choice **d** is incorrect. This expression represents 10 times 2 fewer than the number of video games Joseph owns.

156. **The correct answer is choice b.** To cube means to take the number to the third power. Adding 5 to this yields the expression $x^3 + 5$. Finally, multiplying this by 2 yields $2(x^3 + 5)$.

Choice **a** is incorrect. This represents multiplying the number by 3 as the first step. To cube means to take the number to the third power.

Choice **c** is incorrect. This represents multiplying by 2 before adding 5.

Choice **d** is incorrect. Two times x cubed is not equivalent to x to the 6th power.

157. **The correct answer is choice c.** The original statement can be written as $n + 4 < 5m$. Given the value of m, $5m = 5 \times 6 = 30$; therefore, $n + 4 < 30$. This can be simplified further, to $n < 26$.

Choice **a** is incorrect. The original statement can be written as $n + 4 < 5m$. This statement can be used to show what n is less than, but it can't indicate what n is greater than.

Choice **b** is incorrect. The original statement can be written as $n + 4 < 5m$. Given the value of m, $5m = 5 \times 6 = 30$; therefore, $n + 4 < 30$. While $n < 26$, it is not necessarily true that $n + 4 < 26$.

Choice **d** is incorrect. The original statement can be written as $n + 4 < 5m$. This statement can be used to show what n is less than, but it can't indicate what n is equal to.

158. **The correct answer is choice a.** Call "a number" x. "Times" indicates multiplication. "Less than" is the symbol $<$. So, $5x < 45$ is the correct inequality to represent the given statement.

159. **The correct answer is choice a.** To calculate the cost of renting the car, you need to multiply the number of miles driven (*m*) by $0.50 and add the $45 charge to that answer.

Choice **b** is incorrect because you added the $45 to the miles before finding the amount per mile.

Choice **c** is incorrect because $45 is the initial fee and should not be multiplied by the total number of miles driven.

Choice **d** is incorrect because the values are reversed: $45 is the initial cost and $0.50 should be multiplied by the total number of miles to find the final fee.

160. **The correct answer is choice d.** You can use the following steps to solve this equation:

$$-3x + 6 = 3$$

Subtract 6 from both sides of the equation:

$$-3x = -3$$

Divide both sides of the equation by −3:

$$\tfrac{-3}{-3}x = \tfrac{-3}{-3}$$

Simplify:

$$x = 1$$

To check the solution, substitute $x = 1$ into the equation and see if the equation is true:

$$-3(1) + 6 = 3$$
$$-3 + 6 = 3$$
$$3 = 3$$

Choice **a** is incorrect because if x is −3, then $-3x + 6 = -3(-3) + 6 = 15$, not 3.

Choice **b** is incorrect because if x is −1, then $-3x + 6 = -3(-1) + 6 = 9$, not 3.

Choice **c** is incorrect because if x is 0, then $-3x + 6 = -3(0) + 6 = 6$, not 3.

161. **The correct answer is choice d.** Add like terms:

$$6x + 3x = 9x$$
$$5y - 5y = 0$$
$$9x + 0 = 9x$$

162. **The correct answer is choice b.** Let C = the number of cherries. It is given that 3 apples and 6 oranges equals $\tfrac{1}{2}C$, or $9 = \tfrac{1}{2}C$. Therefore, $C = 2(9) = 18$.

163. **The correct answers are 910 and 1,115.** The minimum number of items the factory could produce in this time frame is $16 \times 8 \times 7 = 896$ items, while the maximum is $20 \times 8 \times 7 = 1{,}120$. Any whole number value in between these numbers is a possible number of items the factory could produce over the given time frame.

164. **The correct answer is choice c.** Subtracting 2 from both sides yields the solution $x > 3$.

Choice **a** is incorrect. In this inequality, the 2 is added to the variable. Therefore, when attempting to isolate the x, both sides should not be multiplied by 2. Instead, 2 should be subtracted from both sides.

Choice **b** is incorrect. In this inequality, the 2 is added to the variable. Therefore when attempting to isolate the x, 2 should be subtracted from both sides instead of being added.

Choice **d** is incorrect. In this inequality, the 2 is added to the variable. Therefore, when attempting to isolate the x, both sides should not be divided by 2. Instead, 2 should be subtracted from both sides.

165. **The correct answer is choice b.** Using the addition method, adding the two equations yields the equation $2x = 22$, which has a solution of $x = 11$.

Choice **a** is incorrect. Subtracting the two equations will eliminate the x from both equations, making it where y must be found first.

Choice **c** is incorrect. If there were infinitely many solutions, the equations would be multiples of each other.

Choice **d** is incorrect. If there were no solution, the equation would yield an incorrect statement such as $0 = 1$ or $-5 = 3$.

166. **The correct answer is choice a.** Distributing the negative and combining like terms yields $(x^2 + 5) - (x^2 - x) = x^2 + 5 - x^2 - (-x) = 5 + x$.

Choice **b** is incorrect. The negative must be distributed to every term in the parentheses.

Choice **c** is incorrect. Since the second term is being subtracted, the x^2 terms will cancel out. Further, the 5 and the x are not being multiplied.

Choice **d** is incorrect. Since the second term is being subtracted, the x^2 terms will cancel out.

167. **The correct answer is 40 $5 blue tickets and 90 $10 red tickets.** We need to use a system of equations to solve this problem. We know that Edward sold 130 red and blue tickets combined. Let's use r for the red tickets and b for the blue tickets. Our first equation is $r + b = 130$. We also know the prices of each ticket and the total cost.

Our second equation is $10r + 5b = 1,100$.

Our system of equations, then, is:

$$r + b = 130$$
$$10r + 5b = 1,100$$

Remember, to solve this problem, we have three possible methods. However, let's use the method of substitution.

If we use the first equation to isolate r, we need to subtract b from both sides of the equation:

$$r + b = 130$$
$$\underline{\quad -b \qquad -b \quad}$$
$$r = 130 - b$$

Now, we can substitute $130 - b$ for r in the second equation.

$$10r + 5b = 1,100$$
$$10(130 - b) + 5b = 1,100$$

Distribute 10 to each of the terms in parentheses.

$$1,300 - 10b + 5b = 1,100$$

Combine the b's and the numbers, each on one side of the equation.

$$1,300 - 10b + 5b = 1,100$$
$$\underline{-1,300 \qquad\qquad\quad -1,300}$$
$$-5b = -200$$

Divide each side by -5 now to isolate b.

$$\frac{-5b}{-5} = \frac{-200}{-5}$$
$$b = 40$$

What was b representing again? The number of blue $5 tickets sold. How are we going to find the number of red tickets? Substitute 40 for b in the equation $r + b = 130$:

$$r + b = 130$$
$$r + 40 = 130$$

Subtract 40 from each side to get r by itself.

$$r + 40 = 130$$
$$\underline{\quad -40 \quad -40 \quad}$$
$$r \quad = \quad 90$$

Therefore, 40 $5 blue tickets were sold and 90 $10 red tickets were sold.

168. The correct answer is choice c. You can use the following steps to solve this equation:

$$4x - 7 = 37$$
$$4x - 7 + 7 = 37 + 7$$
$$4x = 44$$
$$x = 11$$

To check, substitute 11 into the equation and see if the equation is true:

$$4(11) - 7 = 37$$
$$44 - 7 = 37$$
$$37 = 37$$

Choice **a** is incorrect because if x were 7.5, then $4x - 7$ would be 23, not 37.

Choice **b** is incorrect because if x were 10, then $4x - 7$ would be 33, not 37.

Choice **d** is incorrect because if x were 40, then $4x - 7$ would be 153, not 37.

169. The correct answer is choice d. Garrett will always be 2 years younger than Robert. The equation $G = R - 2$ represents this. This year, Robert happens to be twice Garrett's age. Next year, Robert will be 5 and Garrett will be 3 and Robert will no longer be twice Garrett's age. The only relationship that continues for life is that Garrett is 2 years younger than Robert.

170. The correct answer is choice d. Solve the equation for y.

First, isolate the variable by subtracting 11 from each side.

$$2y + 11 - 11 = 25 - 11$$
$$2y = 14$$

Next, divide both sides of the equation by 2:

$$\frac{2y}{2} = \frac{14}{2}$$
$$y = 7$$

Choice **a** is incorrect because you added 11 to 25, instead of subtracting.

Choice **b** is incorrect because you subtracted 2 from 25 and then divided 11 by that value. $2y$ must be treated as its own value, and you must divide by 2 to isolate the y.

Choice **c** is incorrect because you likely made a subtraction error.

171. **The correct answer is choice a.** You can use the following steps
to solve this equation:

$$3z - 2 = 7 + 4z$$

Subtract $3z$ from both sides of the equation:

$$-2 = 7 + z$$

Subtract 7 from both sides of the equation:

$$-9 = z$$

You may have selected choice **b** if you subtracted 7 from the right
side of the equation and added 7 to the left side of the equation,
instead of subtracting 7 from both sides (or if you added 2 to the
left side of the equation and subtracted 2 from the right side of the
equation, instead of adding 2 to both sides). If you selected choice
c, you may have made this same error and also forgotten the negative sign on one side of the equation after combining like terms.

If you correctly subtracted 7 from or added 2 to both sides of the
equation, but forgot the negative sign on one side of the equation
after combining like terms, you may have selected choice **d**.

172. **The correct answer is choice d.** The coat costs $25 more than
the twice the cost of the sweater, so you can let $2s$ represent twice
the cost of the sweater, and then add $25 to represent the cost of
the coat. Therefore, $c = 2s + 25$ is the cost of the coat.

Choice **a** is incorrect because the equation $s = c + 25$ means that the
cost of the sweater is $25 more than the cost of the coat.

Choice **b** is incorrect because the equation $c = s + 25$ means that the
cost of the coat is only $25 more than the cost of the sweater. But
the cost of the coat is $25 more than *twice* the cost of the sweater.

Choice **c** is incorrect because the equation $s = 2c + 25$ means that
the cost of the sweater is $25 more than twice the cost of the coat.
But this does not make sense, because the coat is more expensive
than the sweater.

173. **The correct answer is choice c.** Remember, when you multiply binomials, use the FOIL method.

    F: $(7j)(3j) = 21j^2$

    O: $(7j)(-k) = -7jk$

    I: $(2k)(3j) = 6jk$

    L: $(2k)(-k) = -2k^2$

      $= 21j^2 - 7jk + 6jk - 2k^2$

    Combine like terms:

      $= 21j^2 - jk - 2k^2$

    Choice **a** is incorrect because the middle term was left out.

    Choice **b** is incorrect because $7jk$ was added to $6jk$ instead of being subtracted.

    Choice **d** is incorrect because $6jk$ was subtracted from $-7jk$ instead of being added.

174. **The correct answer is choice d.** Adding polynomials is the same as combining like terms. The exponents of the variables must match, or the terms cannot be added. Rearrange the terms so that like terms are next to each other, and then add (be careful of negatives):

    Original order:

      $(2t^3 - 7t^2 + 5) + (4t^2 + 11t - 9)$

    Move like terms:

      $2t^3 - 7t^2 + 4t^2 + 11t + 5 - 9$

    Add like terms:

      $2t^3 - 3t^2 + 11t - 4$

    Notice that there is no term to add to $2t^3$, nor is there a term to add to $11t$; those terms stay the same.

    Choices **a** and **b** are incorrect because unlike terms were added.

    Choice **c** is incorrect because the sum of +5 and –9 is not +4.

175. **The correct answer is choice b.** The cost of $300 is multiplied by the number of helmets, n, which equals the total of $2,700. Choice **a** would be correct if 300 were in the denominator and 2,700 were in the numerator.

    Choice **c** incorrectly states that the number of helmets, divided by the total amount of money raised, equals the cost of one helmet.

    Choice **d** incorrectly states that the total amount of money raised, when multiplied by the number of helmets purchased, equals the cost of one helmet.

176. **The correct answer is choice c.** The total cost of the party was
$350. The total cost includes the $45 fee that Dave paid plus the
cost of the appetizers equally distributed among 13 people. This
distributed cost is represented by x.
Choice **a** is incorrect because it subtracts the rental fee.
Choice **b** is incorrect because it does not include Dave when calculating the cost of the appetizers per person.
Choice **d** is incorrect because it does not include Dave when calculating the cost of the appetizers per person *and* subtracts the rental
fee.

177. **The correct answer is choice a.** Suewand buys 3 lbs. of almonds
at $8.99 per pound for a total of $26.97. She buys 1.5 lbs. of raisins
at $2.50 per pound for a total of $3.75. So far, she has spent $30.72.
The amount Suewand will spend on chocolate pieces is $3.99x$,
where x = number of pounds. Suewand's total cannot exceed
$37.00, but it can be less than that.
Choice **b** is incorrect because it subtracts instead of adding the
costs of the ingredients together.
Choice **c** is similar to choice **b**, in addition to having the wrong
inequality sign, implying that Suewand can spend more than
$37.00.
Choice **d** correctly adds the costs, but has the wrong inequality sign.

178. **The correct answer is choice b.** First write the system as

$$\begin{cases} y = ax - 3 \\ y = \frac{1}{a}x + \frac{1}{a} \end{cases}$$

The only way for such a system to have no solution is for the slopes
to be equal but the y-intercepts to differ. Here, the slopes are equal
when $a = \frac{1}{a}$, which occurs when $a = -1$ or 1.
Choices **a**, **c**, and **d** are incorrect because the slopes of the lines
are different for these values of a, so the lines will intersect at
least once, thereby giving a solution of the system.

179. **The correct answer is choice a.** The remaining amount to be deposited into the savings account is $6,500 - x$. Multiply each of these amounts by the respective interest rate, add the resulting expressions, and set it equal to $100.50. Doing so yields this expression.

Choice **b** is incorrect because the right side should equal $100.50.

Choice **c** is incorrect because the coefficients on the left side show the interest rates but should show the total of the principal and the interest.

Choice **d** is incorrect because the quantity $x - 6,500$ should be $6,500 - x$. The coefficients on the left side show the interest rates but should show the total of the principal and the interest.

180. **The correct answer is choice a.** Let x be the number of quarters in the collection. There are $x - 2$ nickels and $2x + 1$ dimes. Multiply each quantity by the value of the type of coin, sum these totals, and set it equal to 3.50:

$$0.05(x - 2) + 0.10(2x + 1) + 0.25x = 3.50$$
$$0.50x - 0.10 + 0.10 = 3.50$$
$$x = 7$$

There are $7 - 2 = 5$ nickels.

Choice **b** is incorrect because this is the number of quarters.

Choice **c** is incorrect because this is twice the number of quarters; the number of nickels is 2 less than the number of quarters.

Choice **d** is incorrect because this is the number of dimes.

181. **The correct answer is choice a.** Start by multiplying the two binomials inside the brackets, and then square the resulting quantity: $[(1 - 3x^2)(1 + 3x^2)]^2 = [1 - 9x^4]^2 = 1 - 9x^4 - 9x^4 + 81x^8 = 1 - 18x^4 + 81x^8$.

Choice **b** is incorrect because you forgot to include the middle term (involving the x^4 terms), and the sign of the x^8 term should be positive.

Choice **c** is incorrect because the exponent of the highest-degree term should be 8, not 6.

Choice **d** is incorrect because when you FOIL the binomials inside the brackets, you should get $1 - 9x^4$, not $1 - 6x^4$. Then, when you squared $1 - 6x^4$, you forgot the middle term.

182. **The correct answer is choice b.** Solve the inequality as follows:
$$-x - 3[4(1 - 2x) + 5] > x$$
$$-x - 3[4 - 8x + 5] > x$$
$$-x - 12 + 24x - 15 > x$$
$$23x - 27 > x$$
$$22x > 27$$
$$x > \frac{27}{22}$$

So the solution set is $\{x \mid x > \frac{27}{22}\}$.

Choice **a** is incorrect because you canceled $-x$ on the left side with x on the right side and divided in the wrong manner: If $ax > b$, then $x > \frac{b}{a}$, not $x > \frac{a}{b}$.

Choice **c** is incorrect because you multiplied only the first term in brackets by the -3 outside; you must apply the distributive property to perform this multiplication.

Choice **d** is incorrect because this is the reciprocal of the correct answer.

183. **The correct answer is choice a.** Gather the terms without a y on one side, simplify, and then isolate y:
$$\frac{7}{x} - \frac{3}{y} = \frac{4}{2x}$$
$$\frac{7}{x} - \frac{4}{2x} = \frac{3}{y}$$
$$\frac{14 - 4}{2x} = \frac{3}{y}$$
$$\frac{5}{x} = \frac{3}{y}$$
$$5y = 3x$$
$$y = \frac{3}{5}x$$

Choice **b** is incorrect because you cannot flip fractions of a sum in this manner: $\frac{a}{b} + \frac{c}{d} \neq \frac{b}{a} + \frac{d}{c}$. Get a least common denominator and combine the fractions.

Choice **c** is incorrect because you divided in the wrong manner: if $ax \leq b$, then $x \leq \frac{b}{a}$, not $x \geq \frac{a}{b}$.

Choice **d** is incorrect because you cannot add fractions by simply adding their numerators and denominators; you need to get a least common denominator.

184. **The correct answer is choice c.** Let x = number of turkey sandwiches and y = number of roast beef sandwiches. The first equation of this system is the sum of the numbers of each type of sandwich, which is 12. The second equation gives the total cost as the sum of costs for each type of sandwich bought, which is also correct. Choice **a** is incorrect. The right sides of these equations should be interchanged.

Choice **b** is incorrect because the cost equation (the first one) is wrong; multiply the number of each type of sandwich by the cost of that type of sandwich and add those two quantities.

Choice **d** is incorrect because the right side of the first equation should be multiplied by 100.

185. **The correct answer is choice b.** First rewrite the system in standard form:

$$\begin{cases} 2x - y = 1 \\ x + 2y = 1 \end{cases}$$

Multiply the first equation by 2: $4x - 2y = 2$. Add this equation to the second one to cancel the y terms: $5x = 3$, so $x = \frac{3}{5}$. Now substitute this into the first equation to get $2(\frac{3}{5}) - y = 1$, so $y = \frac{1}{5}$.

So the solution of the system is $x = \frac{3}{5}, y = \frac{1}{5}$.

Choice **a** is incorrect because this satisfies the second equation, but not the first. So it is not a solution of the system.

Choice **c** is incorrect because when you multiplied the first equation of the system by 2 (to eliminate the terms), you forgot to multiply the constant term by 2.

Choice **d** is incorrect because it satisfies the first equation but not the second one. So it is not a solution of the system.

186. **The correct answer is choice c.** FOIL the binomials and combine like terms:

$$(2x^2 + 3)(2 - 4x^2) = (2x^2)(2) - (2x^2)(4x^2) + (3)(2) - (3)(4x^2)$$
$$= 4x^2 - 8x^4 + 6 - 12x^2$$
$$= -8x^4 - 8x^2 + 6$$

Choice **a** is incorrect because $(a + b)(c + d) \neq ac + bd$.

Choice **b** is incorrect because you dropped the middle squared term.

Choice **d** is incorrect because this is the negative of the correct answer.

187. **The correct answer is choice d.** Solve the equation as follows:

$$-2(x + 2) - 1 = 2(2 - x)$$
$$-2x - 4 - 1 = 4 - 2x$$
$$-2x - 5 = 4 - 2x$$
$$-5 = 4$$

Since the resulting statement is false, we conclude that the equation has no solution. Choices **a**, **b**, and **c** are incorrect because substituting these values for x does not yield a true mathematical statement.

188. **The correct answer is choice d.** Substitute the given information into the equation and solve for g:

$$56.1 = \frac{151.47}{g}$$
$$56.1g = 151.47$$
$$g = \frac{151.47}{56.1} = 2.7$$

Choice **a** is incorrect because you should divide 151.47 by 56.1, not subtract 56.1 from it.

Choice **b** is incorrect because you divided in the wrong order.

Choice **c** is incorrect because this is an arithmetic error—the decimal point is in the wrong location.

189. **The correct answer is choice d.** This expression is not equivalent to the others because when simplifying the original expression, this is obtained by inadvertently canceling $6x^2$ with $-3(-2x^2)$.

Choice **a** is an incorrect choice because this is equivalent to the original expression by the commutative property, since the order of the terms being added has been changed.

Choice **b** is an incorrect choice because this is equivalent to the original expression, since $x - 1$ has been factored out and the order of the resulting terms rearranged.

Choice **c** is an incorrect choice because if you multiply the terms and simplify the original expression, you get $14x^2 - 3x - 12$. If you then factor out -3 from the last two terms, you get the expression here.

190. The correct answer is choice d. Let x be the cost of 1 plum and y be the cost of 1 peach. The first equation in this system describes the cost for 5 plums and 6 peaches and the second equation describes the cost of 8 plums and 3 peaches.

Choice **a** is incorrect because you should set up two cost equations, not combine the costs into one equation.

Choice **b** is incorrect because the right sides of the two equations should be interchanged.

Choice **c** is incorrect because the y-terms in the two equations should be interchanged.

191. The correct answer is choice a. The slope of both lines of this system is 2. So the lines are parallel and they have different y-intercepts. Hence, they do not intersect and the system has no solution.

Choice **b** is incorrect because these two lines have the same slope and y-intercept. So the system has infinitely many solutions.

Choice **c** is incorrect because this system is composed of one vertical line and one horizontal line. So it has a unique solution given by $x = 4, y = -1$.

Choice **d** is incorrect because these two equations are equivalent, so the system has infinitely many solutions.

192. The correct answer is choice c. Since the left side is always non-negative and the right side is less than zero, the solution set of this inequality is the set of all real numbers.

Choice **a** is incorrect because the x-values strictly between -2 and 2 do not satisfy this inequality.

Choice **b** is incorrect because positive x-values and x-values less than -5 do not satisfy this inequality.

Choice **d** is incorrect because 3 does not satisfy this inequality.

193. The correct answer is choice c. Add the equations to eliminate the y terms, resulting in the equation $3x = 3a$. So $x = a$. Plug this into one of the equations in the original system to find y. Using the second equation yields $a - y = 2a$, so that $y = -a$. Choice **a** is incorrect because this satisfies the first equation but not the second.

Choice **b** is incorrect because you switched the values of x and y.

Choice **d** is incorrect because when substituting the x-value $x = a$ (which is correct) back into an equation of the original system to find y, you solved that linear equation incorrectly.

194. **The correct answer is choice b.** Note that $P(\text{both}) = 0.47(0.35)$ $= 0.1645$. Now use the addition formula for computing the probability of a union of two events that have common outcomes to conclude that $P(\text{EITHER scores the point OR has one fault}) = 0.35$ $+ 0.47 - 0.1645 = 0.6555$, or about 0.66.

Choice **a** is incorrect because you did not subtract $P(\text{both})$.

Choice **c** is incorrect because you did not account for the probability of having one fault.

Choice **d** is incorrect because you identified the events incorrectly when using the addition formula.

195. **The correct answer is choice d.** Multiply the first equation by -2, and you get the second one. Therefore, these equations are equivalent; every point on the line satisfies both equations in the system, so the system has infinitely many solutions.

Choice **a** is incorrect because this system has no solution. The lines are parallel since they have the same slope 3 and different y-intercepts.

Choice **b** is incorrect because these lines are perpendicular, so the system has a unique solution.

Choice **c** is incorrect because this system has a unique solution since the lines have different slopes.

196. **The correct answer is choice d.** Solve for x as follows:

$$4(3 - 2x) - 2(1 - 3x) = -2$$
$$12 - 8x - 2 + 6x = -2$$
$$10 - 2x = -2$$
$$-2x = -12$$
$$x = 6$$

Choice **a** is incorrect because you did not apply the distributive property correctly. You must multiply both terms of the binomial by the term outside the parentheses.

Choice **b** is incorrect because the solution of a linear equation of the form $ax = b$ is $x = \frac{b}{a}$, not $\frac{a}{b}$.

Choice **c** is incorrect because when solving a linear equation of the form $a + bx = c$, subtract a from both sides; do not add it to both sides.

197. **The correct answer is choice d.** Solve for r, as follows:

$$V = \tfrac{1}{3}\pi r^2 h$$
$$3V = \pi r^2 h$$
$$\frac{3V}{\pi h} = r^2$$
$$\sqrt{\frac{3V}{\pi h}} = r$$

So $r = \sqrt{\frac{3V}{\pi h}}$.

Choice **a** is incorrect because you should be dividing by πh inside the radicand, not multiplying by it.

Choice **b** is incorrect because you should be dividing by πh, not multiplying by it, and you mishandled the square root.

Choice **c** is incorrect because you did not apply the square root correctly. When solving for r, take the square root of both sides instead of multiplying both sides by 2.

198. **The correct answer is choice c.** Let x be the number of vinyl albums. Then the number of CDs is $2x - 3$. Multiply each quantity by the cost per unit and add the resulting expressions to get the total cost $275.50. Doing so results in the equation in choice **c**.

Choice **a** is incorrect because you cannot use two different variables and just one equation if you hope to determine the values of both variables.

Choice **b** is incorrect because the number of CDs is $2x - 3$, not $2x + 3$, where x is the number of vinyl albums.

Choice **d** is incorrect. Let x be the number of vinyl albums. You must multiply the number of CDs by the cost of one CD and the number of vinyl albums by the cost of one vinyl album, rather than combining them into a single term. Also, the number of CDs is $2x - 3$, not $2x + 3$

199. **The correct answer is choice b.** $-x^2(x + 1) - (x^3 + 4x^2) = -x^3 - x^2 - x^3 - 4x^2 = -2x^3 - 5x^2$.

Choice **a** is incorrect. The terms within the second set of parentheses are not like terms and therefore cannot be combined.

Choice **c** is incorrect. The negative must be distributed to every term in the second set of parentheses.

Choice **d** is incorrect. The terms in front of both sets of parentheses must be distributed to every term within the parentheses.

5

Graphs of Linear Equations and Inequalities

Now that you have practiced operating on algebraic expressions, solving linear equations, and modeling word problems with algebra, we'll move on to representing and interpreting them in a coordinate plane. On the GED test you must be able to represent information visually as well as understand some key aspects when interpreting visual relationships. In this chapter you'll also get to test your proficiency with graphing inequalities. Get ready to practice the following concepts:

- Understanding slope as a rate of change

- Calculating the slope from graphs, tables, and coordinate pairs

- Writing linear equations in slope-intercept form: $y = mx + b$

- Graphing linear equations

- Working with parallel and perpendicular lines in the coordinate plane

- Solving systems of linear equations

- Writing linear inequalities to represent real-world situations

- Solving linear inequalities

- Graphing linear inequalities on number lines

200.

x	0	2	4	6
y	1	4	7	10

This table shows some points in the x-y coordinate plane that the graph of a line $y = mx + b$ passes through. Based on this information, what is the value of the slope m?

a. $\frac{1}{2}$

b. $\frac{2}{3}$

c. $\frac{3}{2}$

d. 2

201. Which of the following is the equation of the line that passes through the points $(-8,1)$ and $(4,9)$ in the x-y coordinate plane?

a. $y = \frac{2}{3}x + \frac{19}{3}$

b. $y = \frac{2}{3}x + 9$

c. $y = \frac{3}{2}x + \frac{21}{2}$

d. $y = \frac{3}{2}x + 13$

202. The line n is parallel to the line $y = 3x - 7$ and passes through the point $(5,1)$. At what point does the line n cross the y-axis? Write your answer in the box.

203. A line passes through the point $(4,0)$ and has a slope of $-\frac{1}{2}$. What is the equation of this line?

a. $y = -\frac{1}{2}x + 2$

b. $y = -\frac{1}{2}x - 2$

c. $y = -\frac{1}{2}x + 4$

d. $y = -\frac{1}{2}x - 4$

204. What is the equation of the line that passes through the points
$(-2,1)$ and $(4,5)$ in the Cartesian coordinate plane?

a. $y = \frac{2}{3}x - \frac{4}{3}$

b. $y = \frac{2}{3}x - \frac{1}{3}$

c. $y = \frac{2}{3}x + \frac{7}{3}$

d. $y = \frac{2}{3}x + 4$

205. What is the slope of the line represented by the equation
$10x - y = 2$?

a. 1

b. 2

c. 5

d. 10

206. A line is perpendicular to the line $y = \frac{5}{6}x + 1$ and has a y-intercept
of $(0,-4)$. What is the equation of this line?

a. $y = -4x + 1$

b. $y = \frac{5}{6}x - 4$

c. $y = -\frac{6}{5}x + 1$

d. $y = -\frac{6}{5}x - 4$

207. What is the slope of the following line?

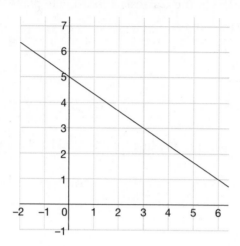

a. $\frac{2}{3}$

b. $-\frac{2}{3}$

c. $-\frac{3}{2}$

d. 5

208. A specimen is removed from an arctic dig and placed in a heating chamber where the temperature is increased steadily over a period of hours. The temperature of the chamber over time is represented on the following graph. What is the temperature in degrees Fahrenheit after 2 hours? _____

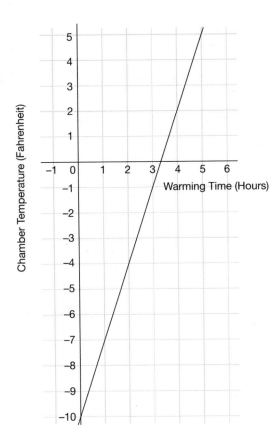

209. Identify the rate of change from the following table:

x	−2	1	7
y	−5	0	10

a. $\frac{5}{3}$

b. $-\frac{5}{3}$

c. $-\frac{3}{5}$

d. $\frac{3}{5}$

210. Which of the following equations has a slope of $\frac{1}{2}$?

a. $2y = \frac{1}{2}x + 10$

b. $y = \frac{1}{2} + 10x$

c. $3x + 6y = 10$

d. $3x - 6y = 10$

211. What is the equation of the line illustrated in the following graph?

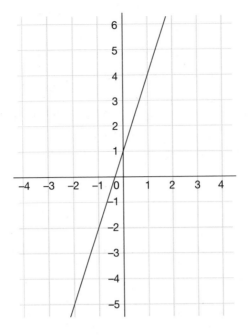

    a. $y = 1x + 3$

    b. $y = \frac{1}{3}x + 1$

    c. $y = 3x - 1$

    d. $y = 3x + 1$

212. What line is parallel to the line $y - 2 = 3x$?

    a. $y = 2x - 1$

    b. $y = 3x + 3$

    c. $y = -2x - 6$

    d. $y = \frac{-1}{3}x + 9$

213. What is the slope of a line perpendicular to $y + \frac{3}{4}x = 1$?

    a. $\frac{-4}{3}$

    b. -1

    c. 1

    d. $\frac{4}{3}$

214. What is an equation of the line that passes through $(-4,3)$ and has a slope of $\frac{1}{2}$?

    a. $x - 2y + 10 = 0$

    b. $2x - 4y - 6 = 0$

    c. $-4x - 3y - 7 = 0$

    d. $-4x + 3y + \frac{1}{2} = 0$

215. Which graph represents two relationships that have the same rate of change?

    a.

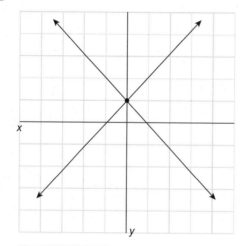

    b.

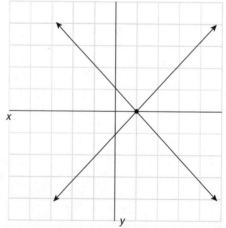

c.

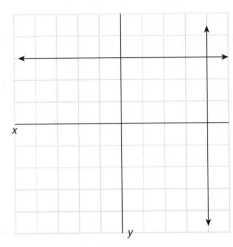

d.

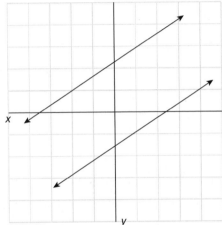

216. What is the solution to the following system of equations?

$y = 3x - 5$ and $2y + 2x = 14$

217. Which number line represents the solution set to the inequality $2x < 24 + 8x$?

a.

-4

b.

-4

c.

-4

d.

-4

218. In the following x-y coordinate plane, draw a dot on the point that is represented by the ordered pair (4,–2).

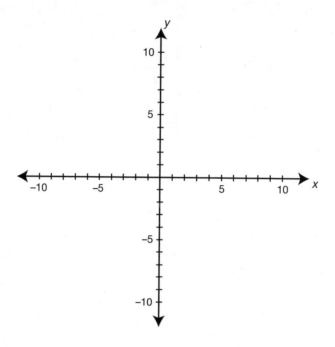

219. A line p passes through the point (–8,4) and has a slope of $\frac{4}{5}$. Which of the following represents the equation for the line p?
   a. $4x - y = -52$
   b. $4x - y = -60$
   c. $4x - 5y = -60$
   d. $4x - 5y = -52$

220. A line P graphed in the x-y coordinate plane crosses the x-axis at a point (–5,0). If another line Q has an equation of $y = 3x - 2$, then which of the following statements is true?
   a. The x-intercept of line P is closer to the origin than the x-intercept of line Q.
   b. The x-coordinate of the x-intercept of line P is smaller than the x-coordinate of the x-intercept of line Q.
   c. The x-intercepts of both lines lie to the right of the y-axis.
   d. The x-intercept of line Q cannot be determined from the given information.

221. Which of the following lines is parallel to the line $y = \frac{2}{9}x - \frac{1}{5}$?

    a. $y = -\frac{9}{2}x + 1$

    b. $y = \frac{3}{4}x + 5$

    c. $y = \frac{2}{9}x - 8$

    d. $y = \frac{3}{4}x - \frac{1}{5}$

222. Draw a dot on this grid to plot the point indicated by the ordered pair (–2,1).

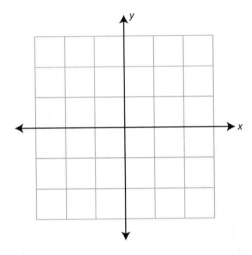

223. The following figure represents the rate of cooling for a particular material after it was placed in a super-cooled bath.

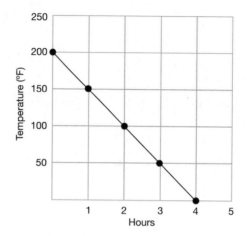

If the temperature, in Fahrenheit, is represented by T and the number of hours elapsed is represented by H, then which of the following would represent a situation where the rate of cooling was faster than the rate indicated in the graph?

a. $T = -25H + 150$

b. $T = -60H + 300$

c. $T = -10H + 200$

d. $T = -50H + 250$

224. The equation of the line that passes through the point $(-2,8)$ and is perpendicular to $y = \frac{1}{4}x + 6$ can be written as $y = \underline{\hspace{1cm}} x + \underline{\hspace{1cm}}$. Fill in the missing values.

225. Circle the line in the following coordinate plane that represents the graph of the equation $3x - 2y = 1$.

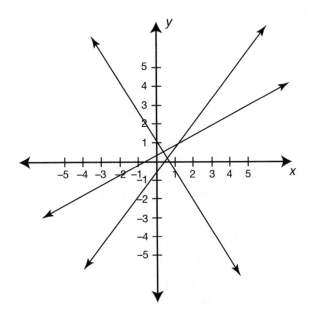

226. A line z is perpendicular to the line $y = -x + 5$. If z passes through the points $(0,-2)$ and $(x,5)$, what is the value of x?

    a. 0

    b. 3

    c. 7

    d. 10

227. Which of the graphs represents the solution for the following systems of equations?

$$\begin{cases} 3(2x + 3y) = 63 \\ \quad\quad 27y = 9(x - 6) \end{cases}$$

a.

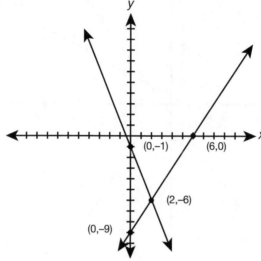

b.

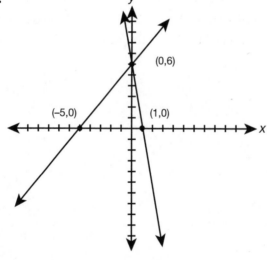

c.

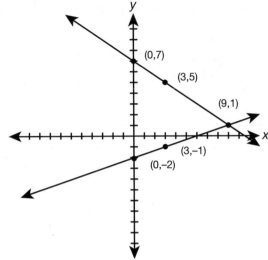

d.

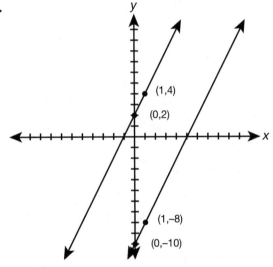

228. The values of x and y in the following table represent a function. What is the rule for this function?

x	y
0	4
1	3
2	2
3	1

a. $y - x = 4$
b. $y + x = 4$
c. $y + 2x = 4$
d. $y - 2x = 4$

229. Which of these equations is shown in the graph?

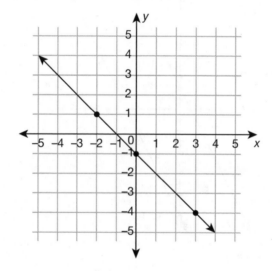

a. $y = x - 1$
b. $y = -x - 1$
c. $y = 2x - 2$
d. $y = 1 + x$

230. Determine the function that best describes the relationship between the x and y values shown in the table:

x	y
−2	8
−1	6
0	4
1	2
2	0

a. $y = x + 6$

b. $y = 2x + 4$

c. $y = -4x$

d. $y = -2x + 4$

231. The graph of the linear function $y = -\frac{1}{2}x + 2$ appears in which one of the following graphs?

a.

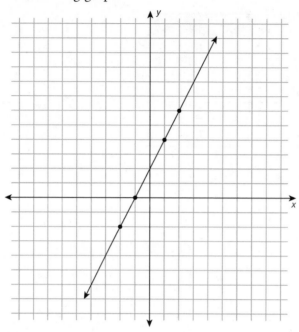

b.

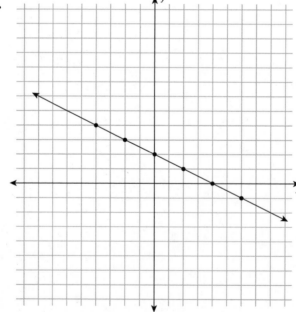

c.

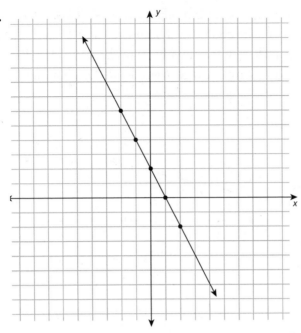

d.

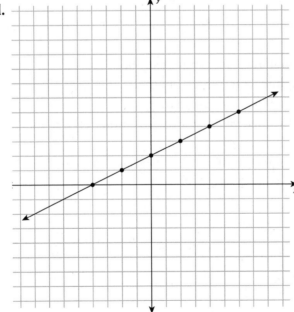

232. The figure represents the cumulative number of packages loaded onto trucks over the course of 8 hours at a small warehouse. When the day began, there were already 50 packages loaded. Based on this graph, how many packages were loaded each hour?

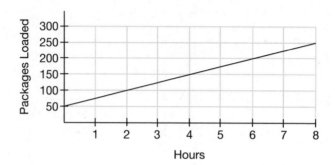

233. What is the slope of the line represented by the equation $10x - 4y = 2$?

234. Which of the following describes the graph of this equation?
$2x - y + 1 = 5 - y$
a. a horizontal line with y-intercept $(0,2)$
b. a vertical line with x-intercept $(2,0)$
c. a horizontal line with x-intercept $(2,0)$
d. a line with slope $m = 2$ that passes through the point $(1,5)$

235. What is the equation of the line that has a slope of $\frac{7}{3}$ and passes through the point $(2,-4)$?
a. $y = \frac{7}{3}x + \frac{26}{3}$
b. $y = \frac{7}{3}x - \frac{26}{3}$
c. $y = \frac{7}{3}x + \frac{14}{3}$
d. $y = \frac{7}{3}x - \frac{14}{3}$

236. Consider the ordered pairs (*x*,*y*) given in the table:

x	-3	2	5	8	11	14
y	7	5	4	-1	-2	-5

Which of the following best describes the line of best fit for this data?

a. The best fit line passes through all of these points.
b. The best fit line is horizontal.
c. The slope of the best fit line is negative.
d. The slope of the best fit line is positive.

237. Which of the following points lies on the line containing the segment shown here?

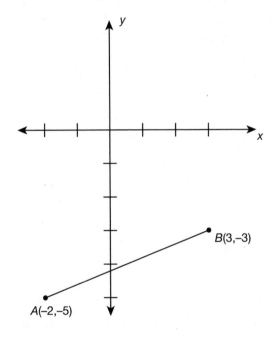

a. (0,–4)
b. (–4,–10)
c. (5,2)
d. (–12,–9)

238. What is the solution set of the inequality $3x - 2(4 - x) < -3x$?

a.

b.

c.

d.

239. Which of the following lines is neither parallel nor perpendicular to the line whose equation is given by $2y + ax = -a$, where a is a positive real number?

a. $y = ax - \frac{a}{2}$

b. $ay = 2x$

c. $ax + 2y = -1$

d. $\frac{a}{2}y = x - a$

240. The slope of line v is $\frac{2}{5}$. If line v goes through points $(-4,7)$ and $(6,x)$, find the value of x. _____

241.

x	0	2	4	6
y	1	4	7	10

This table shows four points in the x-y coordinate plane that lie on the graph of a line $y = mx + b$. Based on this information, what is the value of m? _____

242. The variables x and y are linearly related. When $x = 4$, $y = 6$, and when $x = 10$, $y = 9$. Which equation shows the relationship between x and y?

a. $y = \frac{1}{4}x^2 - 3x + 14$

b. $y = \frac{1}{2}x - 4$

c. $y = 2x + 4$

d. $y = \frac{1}{2}x + 4$

243. What is true about the graph of the linear function $2y = -6$?
   a. The slope of the line is -3.
   b. The x-intercept of the line is -3.
   c. The line is a horizontal line that passes through $(4,-3)$.
   d. The line is a vertical line that passes through $(-3,0)$.

244. What is the equation of the line graphed in this figure?

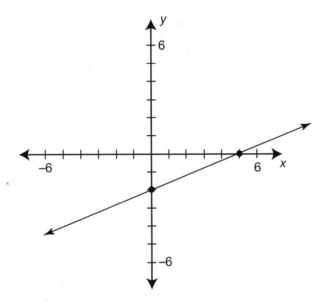

   a. $y = \frac{2}{5}x - 2$
   b. $y = -\frac{2}{5}x - 2$
   c. $y = \frac{2}{5}x + 5$
   d. $y = -\frac{2}{5}x - 5$

Answers and Explanations

200. **The correct answer is choice c.** Using the first two points, $m = \frac{4-1}{2-0} = \frac{3}{2}$.

Choice **a** is incorrect. The slope formula is not $\frac{x_1-y_1}{x_2-y_2}$. In other words, the formula involves subtracting values from different points.

Choice **b** is incorrect. The change in y is represented in the numerator of the slope formula, not the denominator.

Choice **d** is incorrect. The slope formula is not $\frac{x_2-y_2}{x_1-y_1}$. In other words, the formula involves subtracting values from different points.

201. **The correct answer is choice a.** The slope of the line is $m = \frac{9-1}{4-0\,(-8)} = \frac{8}{12} = \frac{2}{3}$. Using this in the point-slope formula along with the first point, the equation can be found with the following steps:

$$y - 1 = \tfrac{2}{3}(x - (-8))$$
$$y - 1 = \tfrac{2}{3}(x + 8)$$
$$y - 1 = \tfrac{2}{3}x + \tfrac{16}{3}$$
$$y = \tfrac{2}{3}x + \tfrac{19}{3}$$

Choice **b** is incorrect. In the point-slope formula, the slope should multiply the entire term $(x - x_1)$.

Choice **c** is incorrect. The slope of the line should be the change in y divided by the change in x. Additionally, only one point should be used in the formula instead of an x-value from one point and a y-value from another.

Choice **d** is incorrect. The slope of the line should be $\frac{2}{3}$, the change in y divided by the change in x.

202. **The correct answer is (0,–14).** Since n is parallel to the given line, it must have the same slope, 3. Given this and the point that n passes through, we can use the point-slope formula to determine the equation for n.

$$y - 1 = 3(x - 5)$$
$$y - 1 = 3x - 15$$
$$y = 3x - 14$$

Now that the equation is in the form $y = mx + b$, we can see that the y-intercept is –14. By definition, this means that the line passes over the y-axis at the point (0,–14).

203. **The correct answer is choice a.** The answer choices are in the form $y = mx + b$. Using the given information, when $x = 4$, $y = 0$, and the slope is $m = -\frac{1}{2}$, this gives the equation $0 = -\frac{1}{2}(4) + b$, which has a solution of $b = 2$.

Choice **b** is incorrect. When solving for the y-intercept b, the -2 must be added to both sides of the equation.

Choice **c** is incorrect. The given point $(4,0)$ is not a y-intercept; it's an x-intercept. The equation $y = mx + b$ uses a y-intercept.

Choice **d** is incorrect. If the x-intercept is $(4,0)$ as given, the y-intercept will be -4 only if the slope is 1. Here the slope is $-\frac{1}{2}$.

204. **The correct answer is choice c.** Using the slope formula first, $m = \frac{5-1}{4-(-2)} = \frac{4}{6} = \frac{2}{3}$. Now, applying the point-slope formula, we have:

$$y - 1 = \frac{2}{3}(x - (-2))$$
$$y - 1 = \frac{2}{3}(x + 2)$$
$$y - 1 = \frac{2}{3}x + \frac{4}{3}$$
$$y = \frac{2}{3}x + \frac{4}{3} + 1 = \frac{2}{3}x + \frac{7}{3}$$

Choice **a** is incorrect. In the point-slope formula, the x_1 and y_1 must come from the same point.

Choice **b** is incorrect. When the point $(-2,1)$ is used in the point-slope formula, the result is $y - 1 = m(x - (-2))$. On the right-hand side of this equation, the 2 ends up being positive.

Choice **d** is incorrect. The slope is found using the change in y on the numerator: $\frac{5-1}{4-(-2)} = \frac{4}{6} = \frac{2}{3}$. In the point-slope formula, $\frac{4}{3} + 1 = \frac{7}{3}$.

205. **The correct answer is choice d.** To find the slope of the line with this equation, move the y-variable to one side on its own to put the equation in the form $y = mx + b$, where m is the slope. Adding y to both sides and subtracting 2 from both sides gives the equation $y = 10x - 2$, so the slope is 10.

Choice **a** is incorrect. The coefficient of x, not the coefficient of y, represents the slope when the equation is written in the form $y = mx + b$.

Choice **b** is incorrect. The slope cannot be read from the equation in the form in which it is currently written.

Choice **c** is incorrect. When solving for y to find the slope, 10 will be divided by 1 and not by 2.

206. **The correct answer is choice d.** The slope will be the negative reciprocal of the given slope, and b in the equation $y = mx + b$ is –4. Choice **a** is incorrect. The slope of a perpendicular line will be the negative reciprocal of the slope of the original line.

Choice **b** is incorrect. Parallel lines have the same slope, while perpendicular lines have negative reciprocal slopes.

Choice **c** is incorrect. The term added to the x term will be the y-intercept, which is not –1.

207. **The correct answer is choice b.** First, select two coordinate pairs that sit on the line. We selected (3,3) and (6,1):

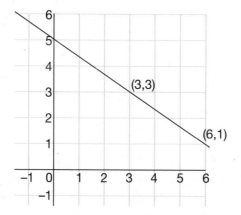

Now put these two coordinates into the slope formula:

$$\text{Slope} = \frac{y_2 - y_1}{x_2 - x_1} = \frac{1 - 3}{6 - 3} = \frac{-2}{3}$$

The slope of this line is $-\frac{2}{3}$, so choice **b** is the correct answer.

Choice **a** is not correct since this shows a positive slope but the line is clearly sloping downward from left to right.

Choice **c** is the incorrect solution to arrive at when the x-values are used in the numerator of the slope formula instead of in the denominator.

Choice **d** shows the y-intercept and not the slope and is therefore incorrect.

208. **The correct answer is –4 degrees.** Looking at the graph, we can see that when $x = 2$ hours, the line has a y-value of –4 degrees:

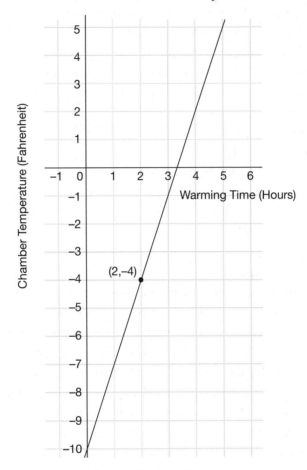

209. **The correct answer is choice a.** The rate of change is the slope of the relationship. Use the coordinate pairs (1,0) and (7,10) in the slope formula to calculate the rate of change:

$$\text{Slope} = \frac{y_2 - y_1}{x_2 - x_1} = \frac{10 - 0}{7 - 1} = \frac{10}{6} = \frac{5}{3}$$

So the rate of change is $\frac{5}{3}$.

Choice **b** is not correct because the slope should be positive.

Choice **d** is the incorrect answer found when the x-coordinates are put in the numerator of the slope formula, and choice **c** is also a result of that type of error.

210. **The correct answer is choice d.** When an equation is in the form $y = mx + b$, the coefficient of x, which is m, always represents the slope. Choice **a** is incorrect because the equation is not yet in $y = mx + b$ form since it begins with $2y$. When everything is divided by 2, the coefficient of x will no longer be $\frac{1}{2}$. The slope in choice **b** is 10 because that is the coefficient x. In choices **c** and **d**, manipulate the equations so that y is by itself. Here we isolate y in choice **d**:

$$3x - 6y = 10$$
$$\underline{-3x \qquad\qquad -3x}$$
$$-6y = -3x + 10$$
$$\frac{-6y}{-6} = \frac{-3x + 10}{-6}$$
$$y = \frac{1}{2}x - \frac{10}{6}$$

So choice **d** gives a slope of $\frac{1}{2}$. Choice **c** is not correct because that will yield a slope of $-\frac{1}{2}$.

211. **The correct answer is choice d.** First, recognize that this line has a y-intercept of 1. That will be the b in $y = mx + b$. Next, you can see that from the point $(0,1)$, you must move 3 spaces *up* and 1 space *over* to reach the point $(1,4)$ on the graph. Since slope is the $\frac{\text{rise}}{\text{run}}$ of a line, we can record this information as slope $= \frac{3}{1}$. So the slope, or m, in the equation will be 3. The final equation will be $y = 3x + 1$, and choice **d** is the correct answer.

Choice **a** is not the correct answer because here the slope and y-intercept have been written in the wrong places.

Choice **b** is incorrect since the slope is not equal to $\frac{1}{3}$, but is instead 3.

Choice **c** is not the correct answer because that has a y-intercept of -1.

212. **The correct answer is choice b.** The slope of the given line is 3, and the slope of the line in choice **b** is also 3. Parallel lines have the same slope. Choices **a** and **c** mistakenly use the y-intercept value of 2, and choice **d** has a slope that renders it perpendicular to the given line.

213. **The correct answer is choice d.** To accurately identify the slope of a line perpendicular to the given one when the x term is on the other side of the equation, manipulate the equation to put it in $y = mx + b$ form.

$$y + \tfrac{3}{4}x = 1$$
$$\underline{-\tfrac{3}{4}x \qquad -\tfrac{3}{4}x}$$
$$y \qquad = \tfrac{-3}{4}x + 1$$

The slope is $\tfrac{-3}{4}$, so the perpendicular slope must be the opposite reciprocal—a positive $\tfrac{4}{3}$.

Choices **b** and **c** confuse the y-intercept term for the slope.

Choice **a** has the wrong sign because the x term was not moved to the other side of the equation before identifying the slope.

An infinite number of lines could be parallel and perpendicular to a given line when we look only at the slope as the determining factor. The only thing that changes with each of these lines is the y-intercept—where it crosses the y-axis.

214. **The correct answer is choice a.** Use the point-slope form of an equation and the information given to answer the question: $y - y_1 = m(x - x_1)$.

Given:

$$x_1 = -4$$
$$y_1 = 3$$
$$m = \tfrac{1}{2}$$

Equation: $y - y_1 = m(x - x_1)$

Substitute: $y - 3 = \tfrac{1}{2}[x - (-4)]$

Simplify: $y - 3 = \tfrac{1}{2}(x + 4)$

Multiply by 2 to clear fractions: $2y - 6 = x + 4$

Add 6 to both sides: $2y = x + 10$

Rearrange terms to look like choices: $0 = x - 2y + 10$

215. **The correct answer is choice d.** Graph **a** shows two relationships that have the same y-intercept, but one rate of change is positive and the other is negative so their rates of change are not equal. Graph **b** shows two relationships that have the same x-intercept, but one rate of change is positive and the other is negative so their rates of change are also not equal. Graph **c** shows two relationships that are perpendicular but their rates of change are not equal. Graph **d** shows two parallel lines, and parallel lines always have an equal rate of change, so this is the correct answer.

216. **The correct answer is (3,4).** Since one of the variables is already isolated in this equation, it can be solved most easily through substitution.

$$y = 3x - 5 \text{ and } 2y + 2x = 14$$

First, replace the y in the second equation with $3x - 5$ from the first equation:

$$2(3x - 5) + 2x = 14$$

Next, solve $2(3x - 5) + 2x = 14$ for x by distributing the 2 and then using the rules of opposite operations to get x alone:

$$
\begin{array}{r}
6x - 10 + 2x = 14 \\
\underline{+10 \qquad\quad +10} \\
8x = 24 \\
x = 3
\end{array}
$$

Last, solve for y by substituting $x = 3$ into one of the original equations:

$$y = 3x - 5, \text{ and } x = 3, \text{ so } y = 3(3) - 5 = 4$$

So (3,4) is the solution to the system of equations.

217. **The correct answer is choice a.** Given $2x < 24 + 8x$, isolate x by using opposite operations.

Subtract $8x$ from both sides to move all the x terms to the left:

$-6x < 24$

Divide by -6 to get x alone, and switch the direction of the inequality sign:

$x > -4$

The graph of this inequality must have an open circle at -4 to show that -4 is not part of the solution set, and it must be shaded to the right to include values greater than -4, so choice **a** is correct. Number line **b** shows the inequality $x \geq -4$. The number lines in graphs **c** and **d** show the inequality $x < -4$ and $x \leq -4$, respectively. In these cases, you forgot to switch the direction of the inequality symbol when dividing by a negative.

218. To graph the point (4,–2), start at the origin (0,0) and move 4 units to the right (since the *x*-coordinate is positive) and 2 units down (since the *y*-coordinate is negative) and then graph the following point:

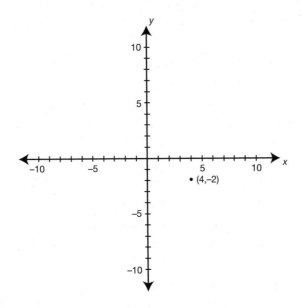

219. **The correct answer is choice d.** Using the point-slope formula, the equation of the line must be

$$y - 4 = \tfrac{4}{5}(x - (-8))$$
$$y - 4 = \tfrac{4}{5}(x + 8)$$
$$y - 4 = \tfrac{4}{5}x + \tfrac{32}{5}$$
$$y = \tfrac{4}{5}x + \tfrac{52}{5}$$

To rewrite this in $Ax + By = C$ form, bring the *x* term to the left-hand side and multiply both sides of the equation by –5.

Choice **a** is incorrect. When rewriting the equation in $Ax + By = C$ form, the –5 must be distributed to all of the terms.

Choice **b** is incorrect. In the point-slope formula, the slope must be distributed to both the *x* term and the constant term. Further, when rewriting the equation in $Ax + By = C$ form, the –5 must be distributed to all of the terms.

Choice **c** is incorrect. In the point-slope formula, the slope must be distributed to both the *x* term and the constant term.

220. **The correct answer is choice b.** The x-coordinate of the x-intercept of line P is -5, while the x-coordinate of the x-intercept of line Q is $\frac{2}{3}$.

Choice **a** is incorrect. The x-intercept of line P is 5 units away from the origin, while the x-intercept of line Q is less than 1 unit away.

Choice **c** is incorrect. The x-coordinate of the x-intercept of line P is negative.

Choice **d** is incorrect. The x-intercept of line Q can be found by letting $y = 0$ and solving for x.

221. **The correct answer is choice c.** This line has the same slope and therefore, by definition, is a line that is parallel to the original.

Choice **a** is incorrect. This line is perpendicular to the given line.

Choice **b** is incorrect. Although the y-intercept is the negative reciprocal of the y-intercept of the original line, this has no effect on whether the line is parallel.

Choice **d** is incorrect. Although the y-intercept is the same as the y-intercept of the original line, this has no effect on whether the line is parallel.

222. To graph the point $(-2,1)$, start at the origin $(0,0)$ and move 2 units to the left (since the x-coordinate is negative) and 1 unit up (since the y-coordinate is positive) and then graph the following point:

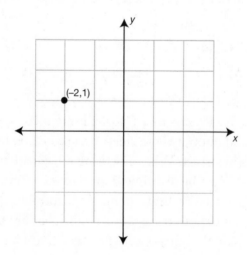

223. **The correct answer is choice b.** The rate of cooling indicated in the graph is the slope of the line passing through the points (0,200) and (4,0). This slope is –50, which implies the material is losing 50 degrees every hour. The slope of the equation in this answer choice is –60, which implies the material is losing 60 degrees every hour, a faster rate of cooling.

Choice **a** is incorrect. This slope would imply that the material is losing 25 degrees every hour, which is a slower rate of cooling.

Choice **c** is incorrect. This slope would imply that the material is losing 10 degrees every hour, which is a slower rate of cooling.

Choice **d** is incorrect. This slope would indicate the material is losing 50 degrees every hour, which is the same rate of cooling that is given in the graph.

224. **The correct answer is $y = -4x + 0$.** If the slope of the given line is $\frac{1}{4}$, what is the slope of any line perpendicular to it? The opposite reciprocal: –4. Now that we have identified the slope, let's substitute the given x and y values into the slope-intercept equation to find the y-intercept of this particular line:

$$y = mx + b$$
$$8 = -4(-2) + b$$
$$8 = 8 + b$$
$$\underline{-8 \quad -8}$$
$$0 = \qquad b$$

The equation of the line perpendicular to $y = \frac{1}{4}x + 6$ that passes through (–2,8) is $y = -4x + 0$, or just $y = -4x$.

225.

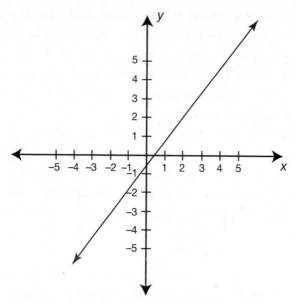

Each of the given lines has a different y-intercept. Solving for y in the given equation will put the equation in $y = mx + b$ form, where b is the y-intercept. In this case, that equation is $y = -\frac{1}{2} + \frac{3}{2}x$. The line given in this image is the only line with a y-intercept of $-\frac{1}{2}$.

226. **The correct answer is choice c.** Since z is perpendicular to $y = -x + 5$, it must have a slope of 1. The given point $(0, -2)$ is a y-intercept since the x-value is 0, so the equation of z must be $y = x - 2$. Plugging in the given y-value of 5 in the point $(x, 5)$ yields the equation $5 = x - 2$, which has the solution $x = 7$.
Choice **a** is incorrect. The y-intercept of the line is -2 and not 5. A perpendicular line does not necessarily have the same y-intercept. Further, the 5 in the point $(x, 5)$ is a y-value and not an x-value.
Choice **b** is incorrect. The 5 in the point $(x, 5)$ is a y-value and not an x-value.
Choice **d** is incorrect. The y-intercept of the line is -2 and not 5. Two perpendicular lines do not necessarily have the same y-intercept.

227. **The correct answer is choice c.** Rewrite the first equation in slope-intercept form:

$$3(2x + 3y) = 63$$

Use the distributive property of multiplication:

$$6x + 9y = 63$$

Subtract $6x$ from both sides:

$$6x - 6x + 9y = 63 - 6x$$

Simplify:

$$9y = 63 - 6x$$

Use the commutative property:

$$9y = -6x + 63$$

Divide both sides by 9:

$$y = \frac{-6}{9}x + 7$$

The equation is in the proper slope-intercept form.

$$m = \frac{-6}{9} = \frac{-2}{3} = \frac{\text{change in } y}{\text{change in } x}$$

$$b = 7$$

The y-intercept is at the point (0,7).

The slope tells you to go down 2 spaces and right 3 for (3,5).

This is the only graph that contains this line.

228. **The correct answer is choice b.** The sum of x and y in each row is 4, so the rule for this function is $y + x = 4$.

Choice **a** is incorrect because when you substitute 1 for x into the equation $y - x = 4$, you get $y = 5$, which is not 3, the corresponding y-value in the table when $x = 1$.

Choice **c** is incorrect because when you substitute 1 for x into the equation $y + 2x = 4$, you get $y = 2$, which is not 3, the corresponding y-value in the table when $x = 1$.

Choice **d** is incorrect because when you substitute 1 for x into the equation $y - 2x = 4$, you get $y = 6$, which is not 3, the corresponding y-value in the table when $x = 1$.

229. **The correct answer is choice b.** The equation of a line is $y = mx + b$, where m is the slope of the line and b is the y-intercept. So, the y-intercept here is -1 and the slope is -1.

Choice **a** is incorrect because it has a positive slope of 1, but the slope of the given line is -1.

Choice **c** is incorrect because the slope of the given line is -1 and the y-intercept is -1, but this equation has a slope of 2 and a y-intercept of -2.

Choice **d** is incorrect because it has a slope of 1, not -1, and a y-intercept of 1, not -1.

230. **The correct answer is choice d.** There are many different ways that you could attempt this problem. You could graph each one of the choices and find out which line contains the values in the table. You could try trial and error by trying to develop your own function. Yet, the easiest method for many people is to simply try each answer until you find the correct function.

Try $y = -2x + 4$.

Does $(-2,8)$ make the equation true?

$8 = -2(-2) + 4$

$8 = 8$ **Yes**

Does $(-1,6)$ make the equation true?

$6 = -2(-1) + 4$

$6 = 6$ **Yes**

Does $(0,4)$ make the equation true?

$4 = -2(0) + 4$

$4 = 4$ **Yes**

Does $(1,2)$ make the equation true?

$2 = -2(1) + 4$

$2 = 2$ **Yes**

Finally, does $(2,0)$ make the equation true?

$0 = -2(2) + 4$

$0 = 0$ **Yes**

Therefore, the answer is $y = -2x + 4$.

231. **The correct answer is choice b.** Use is the slope-intercept method of graphing lines. You have to determine the slope, m, and the y-intercept, b, in order to use this method.

The equation $y = -\frac{1}{2}x + 2$ has a slope of $-\frac{1}{2}$ and a y-intercept of 2. First, graph the point (0,2). Then, to find the other points, move down 1 and to the right 2 because the slope is $-\frac{1}{2}$. In order to get points on the left of the y-axis, move up 1 and to the left 2.

The graph of the equation $y = -\frac{1}{2}x + 2$ looks like the following graph, regardless of the method you use.

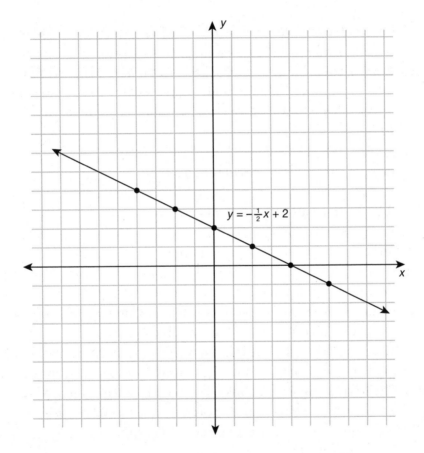

The graph in this choice matches the graph created here.

232. **The correct answer is 25 packages per hour.** The slope of the line will represent the number of packages loaded per hour. Use the starting point (0,50) and the end point (8,250) in the slope formula:

$$m = \frac{y_2 - y_1}{x_2 - x_1}$$
$$m = \frac{250 - 50}{8 - 0} = \frac{200}{8} = 25 \text{ packages per hour}$$

233. **The correct answer is $\frac{5}{2}$.** To find the slope of the line with this equation, isolate the y-variable and put the equation in the form $y = mx + b$, where m is the slope:

$$10x - 4y = 2$$
$$\underline{-10x \qquad\quad -10x}$$
$$-4y = -10x + 2$$
$$\frac{-4y}{-4} = \frac{-10x + 2}{-4}$$
$$y = \frac{-10}{-4}x + \frac{2}{-4}$$
$$y = \frac{5}{2}x + \frac{1}{2}$$

234. **The correct answer is choice b.** To determine what the graph of the equation looks like, first simplify the equation. In general, equations of the form $y = a$ are horizontal lines, equations of the form $x = a$ are vertical lines, and equations of the form $y = mx + b$ are lines that have a slope of m and a y-intercept of $(0,b)$.

Equation:
$$2x - y + 1 = 5 - y$$

Subtract 1 from both sides of the equation:
$$2x - y = 4 - y$$

Add y to both sides of the equation:
$$2x = 4$$

Divide both sides of the equation by 2:
$$x = 2$$

Since the simplified form of the equation is $x = a$, the line is a vertical line that passes through all points of the form $(2,y)$. This is the only choice that is a vertical line, so it is correct.

Remember, once the line is in simplest form, plotting a few points may help. For example, plot the points $(2,0)$, $(2,1)$, $(2,3)$, and so on.

235. **The correct answer is choice b.** To find the equation of a line with a given slope and a given point, use the formula $y = mx + b$. Substitute in the x, y, and m values to solve for b.

$$y = mx + b$$
$$-4 = \frac{7}{3}(2) + b$$
$$-4 = \frac{14}{3} + b$$
$$\frac{-12}{3} = \frac{14}{3} + b$$
$$b = \frac{-26}{3}$$

So, the equation of the line is $\frac{7}{3}x - \frac{26}{3}$.

236. **The correct answer is choice c.** Reading the data values from left to right, the x-values are getting larger while the y-values are getting smaller. So the points fall from left to right, which means the best fit line would do the same. Therefore, the best fit line would have a negative slope.

Choice **a** is incorrect because these points are not all collinear.

Choice **b** is incorrect because reading the data values from left to right, the x-values are getting larger while the y-values are getting smaller. A horizontal best fit line would imply there is little or no change in the y-values.

Choice **d** is incorrect because the best fit line would have a negative slope, not a positive slope.

237. **The correct answer is choice d.** First find the slope of the line containing this line segment:

$$m = \frac{-3 - (-5)}{3 - (-2)} = \frac{2}{5}$$

To get other points on the line, you can add 2 to the y-coordinate and add 5 to the x-coordinate of a point known to be on the line, or you can subtract 2 from the y-coordinate and subtract 5 from the x-coordinate of a point known to be on the line. Using the latter twice in succession gives the points $(-7, -7)$ and $(-12, -9)$.

Choice **a** is incorrect because you seem to have eyeballed the y-intercept. You must use the slope of the line to accurately place points on it.

Choice **b** is incorrect because you subtracted 2 from the x-coordinate and 5 from the y-coordinate of point A but should have done the reverse.

Choice **c** is incorrect because you added 2 to the x-coordinate and 5 to the y-coordinate of point B but should have done the reverse.

238. **The correct answer is choice b.** Solve the inequality as follows:

$$3x - 2(4 - x) < -3x$$
$$5x - 8 < -3x$$
$$8x < 8$$
$$8x < 1$$

This is precisely the set illustrated in this choice.

Choice **a** is incorrect because the endpoint should be open, and the ray should extend to the left, not to the right.

Choice **c** is incorrect because when solving the final inequality $8x < 8$, you seem to have subtracted 8 from both sides instead of dividing. The endpoint of the ray should be 1, not 0.

Choice **d** is incorrect because you should not include the endpoint of the ray, 1, since this is strict inequality.

239. **The correct answer is choice a.** First solve the given equation for y: $y = -\frac{a}{2}x - \frac{a}{2}$.

Its slope is $-\frac{a}{2}$. So any line whose slope is neither $-\frac{a}{2}$ (parallel) nor $\frac{2}{a}$ (perpendicular) is a viable answer. Since the slope of the line in choice **a** is a, it is the correct answer.

Choice **b** is incorrect because the slope of this line is $\frac{2}{a}$, which makes it perpendicular to the given line.

Choice **c** is incorrect because the slope of this line is $-\frac{a}{2}$, which makes it parallel to the given line.

Choice **d** is incorrect because the slope of this line is $\frac{2}{a}$, which makes it perpendicular to the given line.

240. **The correct answer is $x = 11$.** Since the slope is given, and three out of the four coordinates are also provided, plug all of these values into the slope formula and solve for the missing x-coordinate:

$$m = \frac{y_2 - y_1}{x_2 - x_1}$$
$$\frac{2}{5} = \frac{x - 7}{6 - -4}$$
$$\frac{2}{5} = \frac{x - 7}{10}$$

At this point, use cross products to solve for x, making sure you put the quantity $x - 7$ in parentheses so that the 5 gets distributed:

$$5(x - 7) = 2(10)$$
$$5x - 35 = 20$$
$$\underline{+35 \quad +35}$$
$$5x \qquad = 55$$
$$x \qquad = \frac{55}{5}$$
$$x \qquad = 11$$

241. **The correct answer is** $m = \frac{3}{2}$. To answer this question you may use any two coordinate pairs in the slope formula. We will use the first two points (0,1) and (2,4):

$$m = \frac{y_2 - y_1}{x_2 - x_1} = \frac{4 - 1}{2 - 0} = \frac{3}{2}$$

242. **The correct answer is choice d.** It is the only equation that satisfies both conditions. It is a linear equation, and when the values of x and y given in the problem are substituted into the equation, the equation is true.

Choice **a** is incorrect because when the given values of x and y are substituted into the equation, although the result is a true statement, it is not a linear equation.

Choice **b** is incorrect because the substitution of $x = 4$ yields $y = -2$, not 6.

Choice **c** is incorrect because the substitution of $x = 10$ yields $y = 24$, not 9.

243. **The correct answer is choice c.** Solve for y to get a better idea of what the equation looks like.

Equation: $2y = -6$.

Divide both sides by 2: $y = -3$.

Plot a few points and draw a quick sketch of the line. Remember, the points will all look like $(x, -3)$, since $y = -3$.

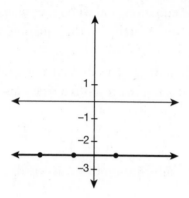

Next, compare the information in this choice to the sketch.

The line is horizontal and passes through all points of the form $(x, -3)$, including $(4, -3)$.

Choice **a** is incorrect because the slope of the line is 0, not -3.

Choice **b** is incorrect because the actual line does not cross the x-axis and therefore it has no x-intercept. (The y-intercept is -3, however.)

Choice **d** is incorrect because the actual line is horizontal, not vertical.

244. **The correct answer is choice a.** Using the two given points, whenever y increases by 2 units, x increases by 5 units. This means the slope must be $m = \frac{2}{5}$ (the change in y divided by the change in x). Further, the y-intercept is $b = -2$. Using the equation $y = mx + b$, we have $y = \frac{2}{5}x - 2$.

Choice **b** is incorrect. The line rises from left to right; therefore, the slope must be positive.

Choice **c** is incorrect. The x-intercept is not used when writing the equation as $y = mx + b$. In fact, b represents the y-intercept.

Choice **d** is incorrect. The line rises from left to right; therefore, the slope must be positive. Additionally, the y-intercept is -2 and not 5.

6

Quadratics and Functions

Quadratic equations are used to model and solve many types of real-world problems that linear equations cannot be applied to since the independent variable is raised to the power of 2. Quadratic equations are used in problems involving finance, gravity, and area. Solving quadratic equations requires a distinct set of skills separate from those used to isolate variables found in linear equations. Functions are special types of relationships that use a unique notation that emphasizes how inputs determine outputs. In this chapter you will be tested on the following skills pertaining to quadratics and functions:

- Standard form of quadratic equations

- Factoring quadratic equations

- The four methods to solve quadratic equations

- Using the quadratic equation

- Interpreting vertices and intercepts of real-world graphs

161

- Recognizing functions

- Working with function notation

- Comparing functions

245. Factor $w^2 - 36$ by dragging and dropping numbers and operations from the following table. (You may use each number or sign more than once.)

$$(w \underline{} \underline{})(w \underline{} \underline{})$$

+	–	1	2	3	4	5	6	7	8	9	16

246. Which of the following functions will have a bell-shaped curve when graphed?
a. $y = \frac{1}{x^2}$
b. $x^3 + 2x^2 + 8 = y$
c. $y = 8x - 3$
d. $(x - 8)^2 = 3x + y$

247. What are the solutions to the equation $17 + x^2 = 81$?
a. 9 and 9
b. $\pm\sqrt{98}$
c. 8 and –8
d. 9 and –9

248. What is the smallest solution to $x^2 - 15x = 100$? _____

249. Find the two solutions to the equation $x^2 - 5x = -6$.
a. 2, –3
b. 2, 3
c. –2, –3
d. –2, 3

250. Without repeating one of the existing coordinate pairs, select two numbers that would make the following a function:

x	f(x)
6	4
5	3
2	8
1	9

1	2	3	4	5	6

251. What function represents the information in the following table?

x	f(x)
−3	−23
0	4
3	−23
6	−104

a. $f(x) = 3x^2 - 4$

b. $f(x) = -2x^2 - 4$

c. $f(x) = 2x^2 - 4$

d. $f(x) = -3x^2 + 4$

252. The graph shown here represents a function $y = g(x)$. What is the vertex of $g(x)$?

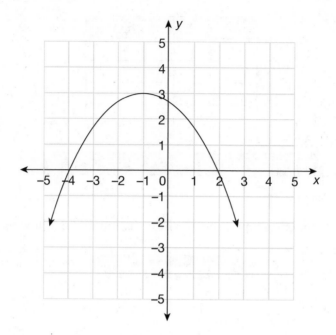

253. Find the two solutions to $2x^2 - 9x - 18 = 0$. _____

254. Which of the following is an equivalent expression of $2y^2 - yp - p^2$?
   a. $(2y + p)(y - p)$
   b. $(y + p)(y - p)$
   c. $2(y^2 - yp - p^2)$
   d. $(2y - p)(y + p)$

255. Where does the function $f(x) = 5x^2 - 25$ intersect the x-axis?
   a. $x = 5, x = -5$
   b. $y = 5, y = -5$
   c. $x = \sqrt{5}, x = -\sqrt{5}$
   d. $y = \sqrt{5}, y = -\sqrt{5}$

256. Which of the following graphs shows *n* as a function of *m*?

a.

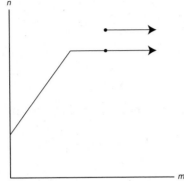

b.

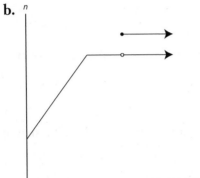

c.

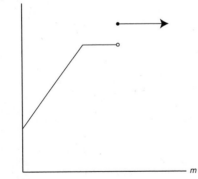

d.

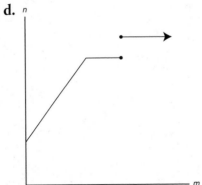

257. For input a, the function f is defined as $f(a) = -2a^2 + 1$. What is the value of $f(-8)$?

   a. -127

   b. -34

   c. 33

   d. 129

258. Which of the following is NOT an equivalent expression to $2x^2 - 2x - 24$?

   a. $2(x^2 - x - 12)$

   b. $2(x - 4)(x + 3)$

   c. $(2x + 6)(x - 4)$

   d. $(x - 3)(2x + 8)$

259. What is the largest possible value of x if $x^2 - 14x + 35 = -10$? Write your answer in the box.

260. The figure shows the graph of a function and all of its turning points.

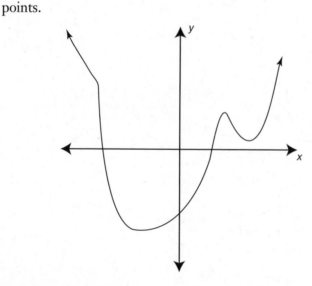

How many x-intercepts does the function have?

   a. none

   b. 1

   c. 2

   d. infinitely many

261. Which of the following is equivalent to the expression $2x + 3(x - 2)^2$?
   a. $3x^2 - 10x + 12$
   b. $3x^2 + 3x - 4$
   c. $3x^2 - 2x + 4$
   d. $3x^2 - 10x + 4$

262.

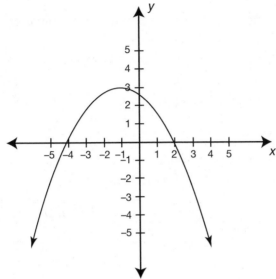

The graph shown here represents a function $y = g(x)$. Select the correct description of the function from the options that follow.
   a. The function has a maximum value of -1 when $x = 3$.
   b. The function has a maximum value of 3 when $x = -1$.
   c. The function has a minimum value of 3 when $x = -1$.
   d. The function has a minimum value of -1 when $x = 3$.

263. What are the two linear factors of the polynomial $2x^2 - x$?
   a. x and $2x - 1$
   b. $2x$ and $x - 1$
   c. $2x$ and x
   d. $2x$ and $x - 2$

264. What is the value of $f(-1)$ if $f(x) = 3(x - 1)^2 + 5$?
   a. 8
   b. 11
   c. 15
   d. 17

265. What is a positive solution to the equation $x^2 - 5x = 14$?
   a. 2
   b. 7
   c. 5
   d. 9

266. Which of the following are the two solutions to the equation $x^2 - 2x - 3 = 0$?
   a. 3 and −1
   b. −3 and 1
   c. −3 and −2
   d. 2 and 2

267. Use the following variables, numbers, and numerical symbols to factor $y^4 - 25$ (terms may be used more than once):

y	y^2	y^3	y^4	1	5	25	+	−

   (   )(   )

268. For what values of x is the function $f(x)$ undefined?
   $f(x) = \frac{3}{x^2 - 3x + 2}$
   a. −1 and −2
   b. 1 and 3
   c. −1 and −3
   d. 1 and 2

269. I. $h(x)$:

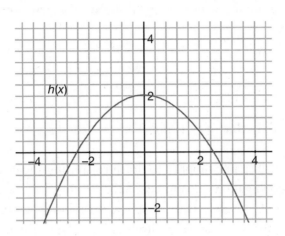

II. $f(x) = -x^2 + 2x$

Given function $h(x)$ displayed in the graph and function $f(x)$ presented algebraically, which phrase correctly completes the following sentence?

The function $h(x)$ has _____ maximum as compared to the maximum of $f(x)$.

a. a greater
b. a smaller
c. an equal
d. an unknown

270. Which of the following shows all numbers excluded from the domain of $g(x) = \frac{x^2 - 4}{9 - x^2}$?

a. 3
b. 2 and –2
c. 9
d. 3 and –3

271. If $x^2 - 11x + 24 = (x + m)(x + n)$, and $m > n$, what is the value of $m - n$?

a. 5
b. 2
c. 11
d. –11

272. Which of the following binomials is NOT a factor of $x^4 - 81$?
   a. $x^2 - 9$
   b. $x - 3$
   c. $x + 9$
   d. $x + 3$

273. What is the value of $f(-1)$ if $f(x) = 3x^2 - 6x + 8$? _____

274. Elise bought paint supplies to finish repainting the exterior of her house. Each gallon of paint cost $27.99, and Elise spent an additional $40 on other paint supplies. If her total before taxes was $263.92, and the cost was written as function f, which function notation represents how many gallons of paint Elise bought?
   a. $f(8)$
   b. $f(27.99)$
   c. $f(40)$
   d. $f(263.92)$

275. Solve for the values of x by factoring the polynomial $2x^2 - 4x = 30$.
   a. $x = 6, x = -5$
   b. $x = -3, x = 5$
   c. $x = -6, x = 5$
   d. $x = 3, x = -5$

276. Use the quadratic formula to solve for x: $3x^2 - 5x = 30$.
   a. $\dfrac{5 + \sqrt{385}}{6}$
   b. $\dfrac{5 - \sqrt{385}}{6}$
   c. $\dfrac{-5 \pm \sqrt{385}}{6}$
   d. $\dfrac{5 \pm \sqrt{385}}{6}$

277. Which graph could represent the information in the table?

x	y
–4	16
–3	9
0	0
3	9
4	16

a.

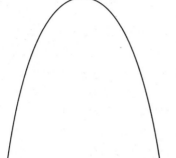

b.

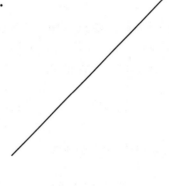

c.

d.

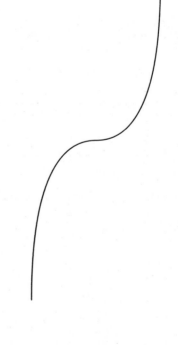

278. According to the table of inputs and outputs, what is the function $f(x)$?

x	3	4	5	6	7
$f(x)$	8	10	12	14	16

a. $f(x) = 3x + 4$
b. $f(x) = 3x - 4$
c. $f(x) = 2x + 2$
d. $f(x) = 2x - 2$

279. The number of bacteria observed in a biological experiment is described by the formula $P(t) = 68(\frac{1}{2})^{-t}$, where $t \geq 0$ is measured in minutes. Which of the following is a true statement?
a. Every minute, the population increases by 68 bacteria.
b. The population levels off after one hour.
c. The population doubles every minute.
d. The maximum number of bacteria is 68.

280. Which of the following number lines correctly illustrates the domain of the function $f(x) = \frac{x}{x-2} + \sqrt{x + 1}$?

a.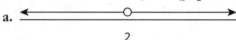

b.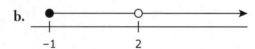

c.

d.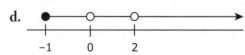

281. For input k, the function f is defined as $f(k) = -2k^2 + 1$. What is the value of $f(-8)$? _____

282. Given the following equation, fill in the chart for all values of $f(x)$.

$$f(x) = 3x^3 - 12$$

x	$f(x)$
1	
3	
5	
7	

283. Consider the following graph of a polynomial function $p(x)$:

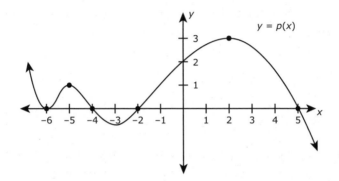

Which of the following is a complete collection of zeros for $p(x)$?

a. $\{-6, -5, -4, -3, -2, 2, 5\}$
b. $\{-4, -2, 5\}$
c. $\{-5, -3, 2\}$
d. $\{-6, -4, -2, 5$

284. Consider the graph of the function :

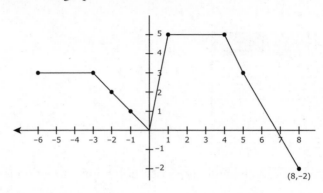

(8,−2)

Over which of the following intervals is the average value of $f(x)$ the largest?
a. (2,5)
b. (−1,1)
c. (−2,1)
d. (−6,5)

285. You measure the amount of rainfall during the spring season. The function $r(x)$ gives the cumulative inches of rainfall, where x is measured in days. Which of the following can be solved to find the number of days it takes to get 7 inches of rain?
a. $r(7) = x$
b. $r(7) = 7$
c. $7r(x) = 1$
d. $r(x) = 7$

286. Concert tickets for a famous cellist go on sale on March 15. The price of a ticket t days after March 15 is given by a function $P(t)$. Which of the following represents the difference in cost of tickets between March 19 and March 21?

a. $P(5) - P(3)$
b. $P(6) - P(4)$
c. $\dfrac{P(4) + P(6)}{2}$
d. $\dfrac{P(19) + P(21)}{2}$

287. $4,200 is deposited into an account that accrues interest continuously at an annual rate of 3%. Which of the following functions gives the amount in the account after t years?

a. $f(t) = 4,200(0.97)^t$
b. $f(t) = 4,200(1.03)^t$
c. $f(t) = 4,200(1.03t)$
d. $f(t) = 1.03(4,200)^t$

288. A popsicle is taken out of the freezer and placed on the counter. Its temperature, T (measured in degrees Fahrenheit) at time t minutes after taking it out of the freezer is given by $T(t) = 60 - 40(\frac{3}{2})^{-t}$. Which of the following statements is/are true?

I. The temperature after two minutes is 40 degrees.
II. $T(t)$ never exceeds 60 degrees.
III. The temperature of a popsicle the moment it is taken out of the freezer is 20 degrees.

a. II only
b. III only
c. II and III only
d. I and III only

289. Factor $x^2 - 12x + 32$ by dragging and dropping numbers and operations from the table. (You may use each number or sign more than once.)

$(x \underline{\ } \underline{\quad})(x \underline{\ } \underline{\quad})$

+	−	1	2	3	4	5	6	7	8	9	16

290. Which of the following graphs could represent the function
$g(x) = x^2 + 4x + 4$?

a.

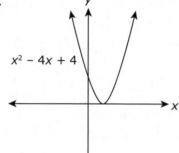

$x^2 - 4x + 4$

b.

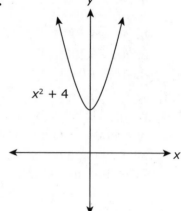

$x^2 + 4$

c.

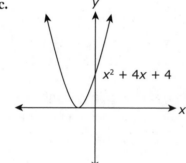

$x^2 + 4x + 4$

d.

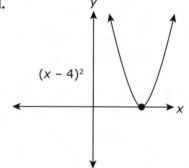

$(x - 4)^2$

291. What represents the complete solutions to $x^2 + 40 = 45$?
  a. 5
  b. 5 and −5
  c. $\sqrt{5}$
  d. $\sqrt{5}$ and $-\sqrt{5}$

292. Identify the smallest value for x that makes the following equation true: $x^2 = 24 - 2x$.

    a. 4

    b. 2

    c. –6

    d. –4

293. What is the largest possible value of x if $x^2 - 14x + 35 = -10$?

    _____

294. Solve the following equation for x: $3x^2 - 11x = 20$.

    a. 5 and $-\frac{3}{4}$

    b. 5 and $-\frac{4}{3}$

    c. –5 and $-\frac{4}{3}$

    d. 5 and $\frac{3}{4}$

295. Theo has a square garden plot in his front yard with side lengths of x feet. He has asked his parents if he can extend the length by 5 feet and the width by 2 feet. His parents are unsure because if he does this expansion his garden will cover 100 square feet of their yard. Which of the following equations could be used to solve for the current dimensions of Theo's garden plot?

    a. $x^2 + 10 = 100$

    b. $4x + 7 = 100$

    c. $(x + 5)(x + 2) = 100$

    d. $(x - 5)(x - 2) = 100$

296. Select the correct response for each underlined section in the following sentence:

When graphed in the coordinate plane, the quadratic equation $10x - 5x^2 + 7 = y$ will open *upward/downward* and will have a y-intercept that is *positive/negative/zero*.

297. Identify the coordinates of the vertex of the following quadratic:

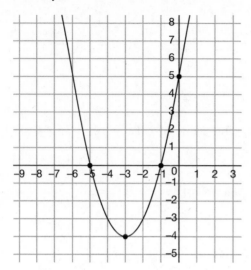

298. Evaluate the function $f(x) = 2x^2 + 5x$ when $x = 3v$.
    a. $36v^2 + 15v$
    b. $51v^3$
    c. $18v^2 + 15v$
    d. $33v^3$

299. Determine whether tables A, B, and C represent functions:

Table A		Table B		Table C	
m	*n*	*p*	*q*	*t*	*s*
2	5	1	5	5	1
3	10	2	5	5	2
1	1	3	5	5	3
0	2	4	5	5	4
3	10	5	5	5	5

Which response demonstrates which of the tables represent functions?
    a. C only
    b. A and B
    c. A and C
    d. A, B, and C

300. If given the function $f(x) = 4x + 10$, what is the value of $f(5) - f(2)$?

    a. 12

    b. 3

    c. 30

    d. 18

301. Determine the function that best describes the relationship between the variables x and y.

x	y
2	3
3	8
4	15
5	24

    a. $y = x^2 - 1$

    b. $y = 2x - 1$

    c. $y = 3x + 1$

    d. $y = 2x + x$

302. Which of the following quadratic equations has two distinct negative real solutions?

    a. $9x^2 = 18x$

    b. $25x^2 + 10x + 1 = 0$

    c. $9x^2 + 9x + 2 = 0$

    d. $x^2 + x + 3 = 0$

303. What is the solution set of the equation $2\sqrt{1 - x} - 10 = 0$?

    a. $\{x \mid x = -26\}$

    b. $\{x \mid x = 16\}$

    c. $\{x \mid x = -24\}$

    d. $\{x \mid x = -4\}$

304. Greg is paid an hourly rate as well as a commission for the number of refrigerators he sells in one day. He gets $13 per hour and works 8-hour days. For every refrigerator he sells, he gets $40. If $g(x)$ is a function that represents the total amount of money Greg earns for selling x refrigerators in a single day, which of the following statements is true?

a. $g(13)$ represents the amount of money Greg earns per hour.

b. $g(8)$ represents the number of hours Greg works in a single day.

c. $g(40)$ represents the number of commission dollars Greg earns for each refrigerator he sells.

d. $g(2)$ represents the amount of money Greg will earn on a day he sells the refrigerators.

Use the following information and illustration to answer questions 305–307.

The graphic shows the flight path of a beanbag that Amara threw from the top of a slide:

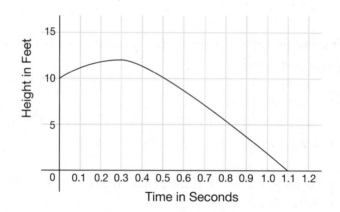

305. From what height was the beanbag thrown? _____

306. After ___ seconds the beanbag was at its peak and was ___ feet from the ground.

307. After how many seconds did the beanbag hit the ground? _____

Answers and Explanations

245. **The correct answer is $(w - 6)(w + 6)$.** The factored form of $w^2 - 36$ will be $(w + m)(w + n)$ where $mn = c$ and $m + n = b$. Since there is no w term, this means that b must equal 0: $w^2 + 0w - 36$. However, $c = -36$. The two factors that multiply to -36 and add to 0 are -6 and 6, so substitute -6 and 6 into $(w + m)(w + n)$: $(w - 6)(w + 6)$.

246. **The correct answer is choice d.** Since quadratics have bell-shaped curves, identify which of the following equations is a quadratic. The equation in choice **a** has an x^2 term in it; however, it is in the bottom of a fraction, so this is not a quadratic. The equation in choice **b** has a leading term of x^3 so this is not a quadratic and is instead a cubic. The equation in choice **c** is in $y = mx + b$ form, so it is therefore a linear equation and will not have a bell-shaped curve when graphed. The equation in choice **d** will have an x^2 term in it on the left-hand side after $(x - 8)^2$ is expanded through FOIL multiplication. This equation is a quadratic that will have a bell-shaped curve.

247. **The correct answer is choice c.** Since there are no x terms in the equation $17 + x^2 = 81$, this is a great opportunity to use the square root technique. Get x^2 alone and then take the square root of both sides:

$$17 + x^2 = 81$$
$$\underline{-17 \qquad -17}$$
$$x^2 = 64$$

Take the square root of both sides:

$$x^2 = 64$$
$$\sqrt{x^2} = \sqrt{64}$$

so $x = 8$ and -8. Incorrect answer choices **a** and **d** were gotten by taking the square root of 81 instead of solving the equation for x. Incorrect answer choice **b** would be the answer arrived at if 81 and 17 were added instead of subtracted.

248. **The correct answer is $x = -5$.** This question requires you to solve $x^2 - 15x = 100$ and identify the larger solution. First rearrange $x^2 - 15x = 100$ so that a is positive and all the terms are on one side of the equation:

$$x^2 - 15x \quad\quad = 100$$
$$\underline{\quad\quad -100 \quad -100}$$
$$x^2 - 15x - 100 = 0$$

The factored form of this will be $(x + m)(x + n) = 0$, where $mn = c$ and $m + n = b$. Identify that $c = -100$ and $b = -15$. Since the two factors must multiply to a negative and add to a negative, we can identify that one factor must be negative and the other must be positive. Create a list of factor pairs that multiply to -100 with the larger factor being negative (that way they will add to a negative b): $\{1,-100\}$, $\{2,-50\}$, $\{4,-25\}$, $\{5,-20\}$, and $\{10,-10\}$. The factors 5 and -20 will multiply to -100 and add to -15, so substitute 5 and -20 into $(x + m)(x + n)$: $(x + 5)(x - 20) = 0$. Now solve for $(x + 5) = 0$ and $(x - 20) = 0$ to arrive at the two solutions of $x = -5$ and $x = 20$. Since the question asked for the *smallest solution*, the correct answer will be $x = -5$.

249. **The correct answer is choice b.** This is a quadratic equation and can be solved in several ways: factoring, completing the square, or using the quadratic formula. However, looking at the terms $-5x$ and -6, it is apparent that factors of 6 can add to 5, so we will factor this quadratic to get the solutions. First, we set the equation equal to 0: $x^2 - 5x + 6 = 0$. We want to find the two values of x where the equation is equal to 0. By factoring, we get $(x - 2)(x - 3)$. Multiply it out if necessary to check that it in fact yields the original equation. Anything multiplied by 0 is 0, so for what values of x would we get an answer of 0? When $x = 2$ and when $x = 3$.

Choices **a**, **c**, and **d** have incorrectly placed positive and negative signs.

250. **The correct answer is 3 or 4.** Since you are not permitted to repeat one of the existing coordinate pairs, you cannot use 6, 5, 2, or 1 as x-values since those are the existing x-values in unique pairings. Only the coordinates 3 or 4 could be chosen for the x value, and then any of the other terms could work for the corresponding y-values.

251. **The correct answer is choice d.** By plugging the values from the table into each of the equations, it is clear that $-3x^2 + 4$ is the only equation that satisfies the relationships between the inputs and outputs. Without even substituting all of the inputs to check the outputs, the input 0 makes it obvious that choice **d** is the only function that satisfies the relationship.

252. **The correct answer is (–1,3).** The vertex of a parabola is its minimum or maximum point. $g(x)$ has a maximum at (–1,3).

253. **The correct answer is $x = 6$ and $x = -\frac{3}{2}$.** Since $a \neq 1$ in $2x^2 - 9x - 18 = 0$, we will solve this using the quadratic formula. First, identify that $a = 2$, $b = –9$, and $c = –18$. Plug these values into the quadratic formula:

$$x = \frac{-b \pm \sqrt{b^2 - 4ac}}{2a}$$

$$x = \frac{-(-9) \pm \sqrt{(-9)^2 - 4(2)(-18)}}{2(2)}$$

$$x = \frac{9 \pm \sqrt{81 + 144}}{4}$$

$$x = \frac{9 \pm \sqrt{225}}{4}$$

$$x = \frac{9 \pm 15}{4}$$

Now solve for $x = \frac{9 + 15}{4}$ and $x = \frac{9 - 15}{4}$:

$$x = \frac{9 + 15}{4} = \frac{24}{4} = 6$$
$$x = \frac{9 - 15}{4} = \frac{-6}{4} = -\frac{3}{2}$$

So, the two solutions are $x = 6$ and $x = -\frac{3}{2}$.

254. **The correct answer is choice a.** Looking at the answer options, we can deduce that we need to factor $2y^2 - yp - p^2$. Keeping in mind how to FOIL backward, we get the two factors $(2y + p)(y - p)$. When these two binomials are multiplied, or FOILed, we get the original expression.

Choice **b** does not have the necessary factor of 2 to get $2y^2$.

Choice **c** incorrectly factors out a 2 from the first term only.

Choice **d** has the addition and subtraction signs incorrectly placed. The way choice **d** is written, we would get $+yp$ instead of $-yp$ when multiplied out.

255. **The correct answer is choice c.** In order to determine where a function hits the x-axis, set $y = 0$ and solve for x:

$$f(x) = 5x^2 - 25$$
$$0 = 5x^2 - 25$$

Since this quadratic doesn't have a b term, solve this equation with the square root technique. To do this, isolate the b^2 and then take the square root of both sides:

$$0 = 5x^2 - 25$$
$$\underline{+25 \qquad\quad +25}$$
$$\frac{25}{5} = \frac{5x^2}{5}$$
$$5 = x^2$$

Now put the x^2 on the left and take the square root of both sides:

$$\sqrt{x^2} = \sqrt{5}$$
$$x = \sqrt{5} \text{ and } x = -\sqrt{5}$$

So choice **c** is the correct answer. Choice **d** mistakenly sets the x-intercept equal to y, but y must have a value of 0 at the point where the curve hits the x-axis. Choices **a** and **b** came about by somehow dropping a 5 and taking the square root of 25.

256. **The correct answer is choice c.** For each possible value of m, there is only one possible value of n.

Choice **a** is incorrect. After and including the indicated point, there are two possible values of n for each value of m.

Choice **b** is incorrect. After the indicated point, there are two possible values of n for each value of m.

Choice **d** is incorrect. At the indicated point, there are two possible values of n for that value of m.

257. **The correct answer is choice a.** $f(-8) = -2(-8)^2 + 1 = -2(64) + 1 = -128 + 1 = -127$.

Choice **b** is incorrect. The exponent on the a indicates a should be squared, not multiplied by 2. Further, the result of this will be positive instead of negative.

Choice **c** is incorrect. The exponent on the a indicates a should be squared, not multiplied by 2.

Choice **d** is incorrect. The value of $(-8)^2$ is positive, not negative.

258. **The correct answer is choice d.** In order to factor $2x^2 - 2x - 24$, first pull the greatest common factor, 2, out of all three terms and rewrite it as $2(x^2 - x - 12)$. This rules out choice **a** as not being an equivalent expression. Next, factor this one step further as $2(x - 4)(x + 3)$ since -4 and $+3$ multiply to -12 and add to -1. This rules out choice **b** as not being an equivalent expression. If the 2 is distributed to the second binomial, $(x + 3)$, the resulting expression is $(x - 4)(2x + 6)$, which is equivalent to the expression in choice **c**. If the 2 is distributed to the first binomial, $(x - 4)$, the resulting expression is $(2x - 8)(x + 3)$, which has signs that are the reverse of those in the expression presented in choice **d**, so choice **d** cannot be an equivalent expression to the original quadratic.

259. **The correct answer is $x = 9$.** Adding 10 to both sides yields the equation $x^2 - 14x + 45 = 0$. The left-hand side of the equation factors into $(x - 5)(x - 9)$, resulting in solutions of 5 and 9. The larger of the two solutions to the equation is 9.

260. **The correct answer is choice c.** The graph crosses the x-axis at exactly two points, and the fact that all of the turning points are shown indicates it will not cross it again.

Choice **a** is incorrect. A graph with no x-intercepts does not cross the x-axis at any point.

Choice **b** is incorrect. A graph with only one x-intercept would cross the x-axis exactly once. This graph crosses the x-axis more than that.

Choice **d** is incorrect. A graph with infinitely many x-intercepts would have to curve back toward the x-axis and cross it in a regular pattern. That behavior is not indicated by this graph since all the turning points are shown.

261. **The correct answer is choice a.** Following the order of operations, the binomial must first be squared, the 3 distributed, and then like terms combined:

$$2x + 3(x - 2)^2 = 2x + 3(x^2 - 4x + 4) = 2x + 3x^2 - 12x + 12$$
$$= 3x^2 - 10x + 12$$

Choice **b** is incorrect. FOIL must be used to expand the squared term: $(x - 2)^2 \neq x^2 + 4$.

Choice **c** is incorrect. When simplifying, the 3 must multiply each term within the parentheses.

Choice **d** is incorrect. When simplifying, the 3 must multiply the constant term 4.

262. **The correct answer is choice b.** The highest point in the graph is the maximum, which is 3. This occurs at $x = -1$. This is the vertex of the parabola.

   Choice **a** is incorrect. The highest point in the graph is the maximum, which is 3. This occurs at $x = -1$. This is the vertex of the parabola.

   Choice **c** is incorrect. The highest point in the graph is the maximum, not the minimum.

   Choice **d** is incorrect. The highest point in the graph is the maximum, which is 3. This occurs at $x = -1$. This is the vertex of the parabola.

263. **The correct answer is choice a.** Both terms share only x as a factor. When this term is factored out, the resulting expression is $x(2x - 1)$.

   Choice **b** is incorrect. The second term does not have a factor of 2, so $2x$ cannot be factored out of the polynomial.

   Choice **c** is incorrect. These two expressions are the factors of the first term. Both are not factors of the second term.

   Choice **d** is incorrect. The second term does not have a factor of 2, so $2x$ cannot be factored out of the polynomial. Further, $x - 2$ is not a factor of the polynomial.

264. **The correct answer is choice d.** Substituting -1 for the x, $f(-1) = 3(-1 - 1)^2 + 5 = 3(-2)^2 + 5 = 3(4) + 5 = 12 + 5 = 17$.

   Choice **a** is incorrect. When substituting -1 for x, $x - 1$ represents $-1 - 1 = -2$, not multiplication.

   Choice **b** is incorrect. It is not true that $(x - 1)^2 = x^2 + 1$.

   Choice **c** is incorrect. By the order of operations, the subtraction within the parentheses as well as the squaring operation must be performed before the multiplication by 3.

265. **The correct answer is choice b.** Rewriting the equation by subtracting 14 from both sides yields the quadratic equation $x^2 - 5x - 14 = 0$. The left-hand side of this equation can be factored into $(x - 7)(x + 2)$, indicating that the solutions are 7 and -2.

   Choice **a** is incorrect. Once the quadratic equation is rewritten and factored, the zero product rule states that $x - 7 = 0$ or $x + 2 = 0$. Therefore one of the solutions is -2 instead of 2.

   Choices **c** and **d** are incorrect. To factor the rewritten quadratic equation, find factors of 14 that sum to -5 instead of numbers that sum to -14.

266. **The correct answer is choice a.** The equation can be factored and rewritten as $(x - 3)(x + 1) = 0$. Using the zero product rule, this results in the equations $x - 3 = 0$ and $x + 1 = 0$. The solutions to these equations are 3 and –1, respectively.

Choice **b** is incorrect. After factoring, the zero product rule must be applied. This will result in the equations $x - 3 = 0$ and $x + 1 = 0$. Choices **c** and **d** are incorrect. The solutions can't be read off the coefficients. Instead, factoring, the quadratic formula, or completing the square should be used to solve a quadratic equation like this.

267. **The correct answer is $(y^2 + 5)(y^2 - 5)$.** The expression $y^4 - 25$ is a difference of perfect squares since both y^4 and 25 are perfect squares. Binomials that are a difference of perfect squares can be factored in the following manner: $a^2 - b^2 = (a + b)(a - b)$; therefore, $y^4 - 25 = (y^2 + 5)(y^2 - 5)$.

268. **The correct answer is choice d.** The function $f(x)$ will be undefined for any x-values that cause its denominator, $x^2 - 3x + 2$, to equal zero. Set this quadratic equal to zero and solve for the x-values by using factoring:

$$x^2 - 3x + 2 = 0$$
$$(x - 2)(x - 1) = 0$$

So when $x = 2$ or when $x = 1$, the denominator of $f(x)$ will equal zero and the function will be undefined.

Choice **a** is incorrect because you forgot to solve for the x-values for each of the binomials in $(x - 2)(x - 1) = 0$ and instead chose the values that were used for factoring.

Choices **b** and **c** are incorrect because $(x - 1)(x - 3)$ and $(x + 1)(x + 3)$ are both incorrect factorizations of $x^2 - 3x + 2$.

269. **The correct answer is choice d.** The maximum of a downward-facing parabola is at its vertex. Looking at the graph of $h(x)$, you can see that it has a maximum of 2. To find the vertex of $f(x) = -x^2 + 2x$, first find the line of symmetry using the formula $x = -\frac{b}{2a}$: $x = -\frac{2}{2(-1)}$, so $x = 1$ is the axis of symmetry. Since the vertex sits on the axis of symmetry, plug $x = 1$ into $f(x)$ to find the y-value of the vertex: $f(1) = -(1)^2 + 2(1) = 1$. So the vertex of $f(x)$ is $(1,1)$, and $h(x)$ has a greater maximum compared to $f(x)$.

270. **The correct answer is choice d.** Any values of x that cause the denominator to equal zero will be excluded from the domain. Set the denominator equal to zero and solve for the x-values that will result in a denominator of 0:

$$9 - x^2 = 0$$
$$9 = x^2$$
$$x = \pm 3$$

So the values of $x = 3$ and $x = -3$ must be excluded from the domain.

Choice **a** is incorrect because it does not show *all* numbers excluded from the domain. It is true that 3 is excluded from the domain, but so is -3.

Choice **b** is incorrect because you set the numerator equal to zero, but you need to set the denominator equal to zero in order to determine what numbers are not in the domain.

Choice **c** is incorrect because you did not take the exponent of 2 into consideration when setting $9 - x^2$ equal to zero.

271. **The correct answer is choice a.** First, factor $x^2 - 11x + 24$ by identifying that -8 and -3 will multiply to 24 and add to -11 and therefore $x^2 - 11x + 24 = (x - 8)(x - 3)$. Since $m > n$, this determines that $m = -3$ and $n = -8$. This means that $m - n = -3 - (-8) = 5$.

Choices **b** and **c** are incorrect because you incorrectly factored the original quadratic as $(x - 8)(x + 3)$, but this is equivalent to $x^2 - 5x + 24$.

Choice **d** is incorrect because you thought that, in the standard form of a factored quadratic, $ax^2 + bx + c = (x + m)(x + n)$, $b = m - n$; however, $b = m + n$.

272. **The correct answer is choice c.** Since both x^4 and 81 are perfect squares, the expression $x^4 - 81$ is a difference of perfect squares. A difference of perfect squares binomial can be factored as follows: $a^2 - b^2 = (a + b)(a - b)$. Therefore, $x^4 - 81 = (x^2 - 9)(x^2 + 9)$.

This shows that choice **a** cannot be the correct answer choice since $x^2 - 9$ is a factor of $x^4 - 81$. Apply the difference of perfect squares factoring technique again to $(x^2 - 9)$:

$$(x^2 - 9)(x^2 + 9) = (x + 3)(x - 3)(x^2 + 9)$$

This rules out answer choices **b** and **d** since $(x + 3)$ and $(x - 3)$ are both factors of $x^4 - 81$. The only given expression that is not a factor is $(x + 9)$.

273. **The correct answer is $f(-1) = 17$.** In order to find the value of $f(-1)$ if $f(x) = 3x^2 - 6x + 8$, replace all of the x-values with -1 and evaluate the expression:

$$f(x) = 3x^2 - 6x + 8$$
$$f(-1) = 3(-1)^2 - 6(-1) + 8$$
$$f(-1) = 3(1) - (-6) + 8$$
$$f(-1) = 3 + 6 + 8$$
$$f(-1) = 17$$

274. **The correct answer is choice a.** This problem can be illustrated as a function in the following way: $f(g) = 27.99g + 40$. The total bill, before taxes, is given, which can be substituted for $f(g)$: $263.92 = 27.99g + 40$. Solve for g:

$$263.92 = 27.99g + 40$$
$$223.92 = 27.99g$$
$$8 = g$$

So, the function notation is $f(8)$.

Choice **b** is incorrect because it inserts the cost of each gallon, not the number of gallons.

Choice **c** is incorrect because it inserts the cost of the extra supplies, not the number of gallons.

Choice **d** is incorrect because it inserts the total cost, not the number of gallons.

275. **The correct answer is choice b.** First, move 30 to the other side so the equation is equal to 0.

$$
\begin{array}{rl}
2x^2 - 4x & = 30 \\
-30 & -30 \\
\hline
2x^2 - 4x - 30 = & 0
\end{array}
$$

Next, factor this quadratic: $(2x + 6)(x - 5)$. Set each factor equal to 0 to find which values of x make this polynomial 0.

$$
\begin{array}{rl}
2x + 6 = & 0 \\
-6 & -6 \\
\hline
2x & = -6 \\
2 & 2 \\
x & = -3
\end{array}
$$

So, one value is $x = -3$.

$$
\begin{array}{rl}
x - 5 = & 0 \\
+5 & +5 \\
x & = 5
\end{array}
$$

The second value is $x = 5$.

276. **The correct answer is choice d.** First, set the quadratic equation equal to 0:

$$3x^2 - 5x \qquad = 30$$
$$\underline{ -30 \qquad -30}$$
$$3x^2 - 5x - 30 = \quad 0$$

Next, identify the a, b, and c values to substitute into the quadratic formula: $a = 3$, $b = -5$, $c = -30$. The quadratic formula is $x = \frac{-b \pm \sqrt{b^2 - 4ac}}{2a}$.

$$x = \frac{-(-5) \pm \sqrt{(-5)^2 - 4(3)(-30)}}{2(3)} = \frac{5 \pm \sqrt{25 - (-360)}}{6} = \frac{5 \pm \sqrt{385}}{6}$$

277. **The correct answer is choice c.** The x-values in the table are squared to get the y-values. The shape of a graph of x^2 is in the shape of a parabola.

Choice **a** is incorrect because a downward parabola results from $y = -x^2$, and the opposite of x^2 is not indicated in the table of values.

Choice **b** is incorrect because it is a linear graph.

Choice **d** is incorrect because it is the graph of a cubic function.

278. **The correct answer is choice c.** By substituting the values in the x row into the equations, it becomes clear that the input, x, is doubled and then increased by 2.

Choices **a** and **b** are incorrect because they yield too big an output since the input is first multiplied by 3.

Choice **d** is incorrect because it decreases by 2 instead of increasing by 2.

279. **The correct answer is choice c.** Observe that $(\frac{1}{2})^{-t} = 2^t$, and so, with every increase in t by one minute, the population doubles.

Choice **a** is incorrect because 68 is the initial population only and does not affect the actual growth rate; here, the population doubles every minute.

Choice **b** is incorrect because this is an example of exponential growth.

Choice **d** is incorrect because, while the population size starts at 68, it continues to increase with time, so this is not the maximum size.

280. **The correct answer is choice b.** The number 2 must be excluded from the domain because it makes the denominator of the first term equal to zero, so that it would be undefined. The radicand must be nonnegative, so that $x \geq -1$. Taking those two conditions together results in the number line in this choice.

Choice **a** is incorrect because x must be greater than or equal to -1, and the radicand must be nonnegative.

Choice **c** is incorrect because you need to exclude 2 from the domain, since it makes the first term undefined.

Choice **d** is incorrect because you should not exclude 0 from the domain because it makes the numerator equal to zero, which is permissible

281. **The correct answer is $f(-8) = -127$.** In order to find the value of $f(-8)$ if $f(k) = -2k^2 + 1$, replace k with -8 and evaluate:

$$f(k) = -2k^2 + 1$$
$$f(-8) = -2(-8)^2 + 1$$
$$f(-8) = -2(-8)(-8) + 1$$
$$f(-8) = -2(64) + 1$$
$$f(-8) = -127$$

282.

x	$f(x)$
1	–9
3	69
5	363
7	1,017

To find the output values, we need to substitute our inputs into the function $f(x)$.

$$f(x) = 3x^3 - 12$$
$$f(1) = 3(1)^3 - 12$$
$$= 3 - 12$$
$$= -9$$
$$f(3) = 3(3)^3 - 12$$
$$= 3(27) - 12$$
$$= 81 - 12$$
$$= 69$$
$$f(5) = 3(5)^3 - 12$$
$$= 3(125) - 12$$
$$= 375 - 12$$
$$= 363$$
$$f(7) = 3(7)x^3 - 12$$
$$= 3(343) - 12$$
$$= 1,029 - 12$$
$$= 1,017$$

Notice that $f(x)$ is the y-coordinate. These points could be plotted on a graph to see this portion of the function.

283. **The correct answer is choice d.** These are the only x-values of points where the graph crosses or touches the x-axis. The x-coordinate of every such point is included in this set. So it is a complete set of zeros.

Choice **a** is incorrect. You should not include in this set those x-values at which there is a maximum or a minimum value of the graph that does not sit on the x-axis.

Choice **b** is incorrect because since the graph touches the x-axis at –6, –6 is a zero of the function.

Choice **c** is incorrect because these are the x-values at which the graph has a maximum or a minimum, not the zeros.

284. **The correct answer is choice b.** The average value of a function on a given interval (a,b) is $\frac{f(b)-f(a)}{b-a}$. Applying this gives $\frac{5-1}{1-(-1)} = 2$, which is the largest of all average values listed in the other choices.

Choice **a** is incorrect because this average value is negative, which is smaller than $(-1,1)$.

Choice **c** is incorrect because this average value is 1, which is smaller than $(-1,1)$.

Choice **d** is incorrect because this average value is 0, which is smaller than $(-1,1)$.

285. **The correct answer is choice d.** Let x = number of days and $r(x)$ = number of inches of rain in 7 days. Then the equation $r(x) = 7$ would be solved for x to answer the posed question.

Choice **a** is incorrect because $r(7)$ = number of inches of rain after 7 days.

Choice **b** is incorrect because this means that after 7 days, you would have 7 inches of rain. This need not be the case.

Choice **c** is incorrect because the 7 should be on the right side, not multiplied by $r(x)$.

286. **The correct answer is choice b.** Here, $t = 4$ corresponds to March 19 and $t = 6$ corresponds to March 21. The difference in the corresponding functional values, namely $P(6) - P(4)$, is the difference in price between these two dates.

Choice **a** is incorrect because each input is off by 1; here, $t = 4$ (not $t = 3$) would correspond to March 19 and $t = 6$ (not $t = 5$) would correspond to March 21.

Choice **c** is incorrect because this is the average price of a ticket for these two dates.

Choice **d** is incorrect because this is an average price of a ticket, but not for the dates shown—you must convert the actual date to the correct t-value.

287. **The correct answer is choice b.** The correct formula to use is $P_0(1 + 4)^t$, where P_0 is the amount invested and r is the interest rate. Using $P_0 = \$4,200$ and $r = 0.03$ yields the formula in choice **b**.

Choice **a** is incorrect because this would mean the value went down 3% each year, not up.

Choice **c** is incorrect because the term $1.03t$ should be $(1.03)^t$.

Choice **d** is incorrect because the 4,200 and 1.03 must be interchanged.

288. **The correct answer is choice c.** Since the term $40(\frac{3}{2})^{-t}$ is positive for all values of t, and it is being subtracted from 60, it follows that $T(t)$ must always be less than 60. So II is true. III is also true because when you evaluate $T(0)$, you get $60 - 40 = 20$.

Choice **a** is incorrect because III is also true because when you evaluate $T(0)$, you get $60 - 40 = 20$.

Choice **b** is incorrect because II is also true. Since the term $40(\frac{3}{2})^{-t}$ is positive for all values of t, and it is being subtracted from 60, it follows that $T(t)$ must always be less than 60.

Choice **d** is incorrect because I is false because when you evaluate $T(2)$, you do not get 40. Also, II is true. Since the term $40(\frac{3}{2})^{-t}$ is positive for all values of t, and it is being subtracted from 60, it follows that $T(t)$ must always be less than 60.

289. **The correct answer is $(x - 8)(x - 4)$.** The factored form of $x^2 - 12x + 32$ will be $(x + m)(x + n)$ where $mn = c$ and $m + n = b$. Identify that $c = 32$ and $b = -12$ in the given quadratic. Since the two factors must multiply to a positive 32 and add to a negative 12, we can identify that both factors must be negative. Come up with a list of factor pairs: $\{-1,-21\}$, $\{-2,-16\}$, and $\{-4,-8\}$. The factors -8 and -4 will multiply to 32 and add to -12, so substitute -8 and -4 into $(x + m)(x + n)$: $(x - 8)(x - 4)$.

290. **The correct answer is choice c.** In order to see where $g(x) = x^2 + 4x + 4$ will have its x-intercepts, factor $g(x)$ into $g(x) = (x + 2)(x + 2)$. This shows that when $x = -2$, the y-value of the function will equal zero. Therefore, the x-intercept is $(-2,0)$. This rules out choice **b** since this graph does not have any x-intercept. This also rules out choices **a** and **d** since both of these graphs have x-intercepts that are positive. The only possible graph to fit the given equation is that illustrated by choice **c** since that has a single, negative x-intercept.

291. **The correct answer is choice d.** Since the equation $x^2 + 40 = 45$ only has an x^2 and does not have an x term, use the square root method. Isolate the x^2 and take the square root of both sides:

$$x^2 + 40 = 45$$
$$x^2 = 5$$
$$\sqrt{x^2} = \sqrt{5}$$
$$x = \sqrt{5} \text{ and } x = -\sqrt{5}$$

Choices **a** and **b** are incorrect because you arrived at $x^2 + 5$, but forgot to take the square root of the x^2.

Choice **c** is incorrect because you forgot that, when solving the quadratic equation $x^2 + 5$, you must include both the negative and positive square roots of 5.

292. **The correct answer is choice c.** First rearrange $x^2 = 24 - 2x$ so that a is positive, all the terms are on one side of the equation, and it looks like $ax^2 + bx + c = 0$.

$$\begin{array}{rcl} x^2 & = & 24 - 2x \\ -24 & & -24 \\ \hline x^2 - 24 & = & -2x \\ +2x & & +2x \\ \hline x^2 + 2x - 24 & = & 0 \end{array}$$

The factored form of this will be $(x + m)(x + n) = 0$ where $mn = c$ and $m + n = b$. Identify that $c = -24$ and $b = 2$. Since the two factors must multiply to a -24 and add to a positive 2, we can identify that one factor will be negative and another factor must be positive. Create a list of factor pairs where the larger factor is positive (that way they will add to a positive b): $\{-1,24\}$, $\{-2,12\}$, $\{-3,8\}$, and $\{-4,6\}$. The factors -4 and 6 will multiply to -24 and add to 2, so substitute -4 and 6 into $(x + m)(x + n)$: $(x - 4)(x + 6) = 0$. Now solve for $(x - 4) = 0$ and $(x + 6) = 0$ to arrive at the two solutions of $x = 4$ and $x = -6$. Out of these two possible values for x, -6 is the smaller value.

Choice **a** is incorrect because although 4 is a solution to the equation, it is not the smaller value to make it true.

Choice **b** is incorrect because $(x - 2)$ is not a factor for the given equation.

Choice **d** is incorrect because you took the factor $(x - 4)$ to mean that -4 would make the equation true; however, this factor represents that 4 is a solution to the equation, since it must be set to zero: $x - 4 = 0$.

293. **The correct answer is $x = 9$.** This question requires you to solve $x^2 - 14x + 35 = -10$ and identify the largest solution. First rearrange $x^2 - 14x + 35 = -10$ so that a is positive, all the terms are on one side of the equation, and it looks like $ax^2 + bx + c = 0$.

$$x^2 - 14x + 35 = -10$$
$$\underline{ +10 \quad +10}$$
$$x^2 - 14x + 45 = 0$$

The factored form of this will be $(x + m)(x + n) = 0$ where $mn = c$ and $m + n = b$. Identify that $c = 45$ and $b = -14$. Since the two factors must multiply to a positive 45 and add to -14, we can identify that both factors must be negative. Create a list of factor pairs that multiply to 45: $\{-1,-45\}$, $\{-3,-15\}$, and $\{-5,-9\}$. The factors -5 and -9 will multiply to 45 and add to -14, so substitute -5 and -9 into $(x + m)(x + n)$: $(x - 5)(x - 9) = 0$. Now solve for $(x - 5) = 0$ and $(x - 9) = 0$ to arrive at the two solutions of $x = 5$ and $x = 9$. Since the question asked for the *largest* possible value of x, the correct answer will be $x = 9$.

294. The correct answer is choice b. First rearrange $3x^2 - 11x = 20$ so that a is positive, all the terms are on one side of the equation, and it looks like $ax^2 + bx + c = 0$:

$$
\begin{array}{rl}
3x^2 - 11x & = 20 \\
\underline{ -20 -20} & \\
3x^2 - 11x - 20 = & 0
\end{array}
$$

Now identify that $a = 3$, $b = -11$, and $c = -20$. Plug these values into the quadratic formula:

$$x = \frac{-b \pm \sqrt{b^2 - 4ac}}{2a}$$

$$x = \frac{-(-11) \pm \sqrt{(-11)^2 - 4(3)(-20)}}{2(3)}$$

$$x = \frac{11 \pm \sqrt{121 + 240}}{6}$$

$$x = \frac{11 \pm \sqrt{361}}{6}$$

$$x = \frac{11 \pm 19}{6}$$

Now solve for $x = \frac{11 + 19}{6}$ and $x = \frac{11 - 19}{6}$:

$$x = \frac{11 + 19}{6} = \frac{30}{6} = 5$$

$$x = \frac{11 - 19}{6} = \frac{-8}{6} = -\frac{4}{3}$$

So, the two solutions are $x = 5$ and $x = -\frac{4}{3}$.

Choice **a** is incorrect because although 5 is a correct solution, $-\frac{3}{4}$ is the reciprocal of the second correct solution, $-\frac{4}{3}$.

Choice **c** is incorrect because although $-\frac{4}{3}$ is a correct solution, -5 is the opposite of the second correct solution, 5.

Choice **d** is incorrect because although 5 is a correct solution, $\frac{3}{4}$ is the negative reciprocal of the second correct solution, $-\frac{4}{3}$.

295. **The correct answer is choice c.** Theo's current garden plot is a square with side lengths of x feet. Since he wants to extend the length by 5 feet and the width by 2 feet, we can write:

new length = $x + 5$

new width = $x + 2$

With these dimensions, the new garden plot would have an area of 100 square feet. Use this information in the area formula to write a quadratic equation:

Area = length × width

$100 = (x + 5)(x + 2)$

Use FOIL to multiply the right side of the equation to get $x^2 + 7x + 10$:

$100 = x^2 + 7x + 10$

This equation could be solved to find the current dimensions of Theo's square garden plot with side length x.

Choice **a** is incorrect because you took the square footage of the current garden (x^2) and added that to 5 × 2, but this doesn't work because Theo isn't just adding an extra 5' × 2' plot onto the garden, but will be increasing the side lengths by these dimensions.

Choice **b** is incorrect because you added the current perimeter ($4x$) with the linear feet to be added onto one length and one width, but this cannot be related to the area to form an equation.

Choice **d** is incorrect because it models what the new area would be if 5 feet were subtracted from the length and 2 feet were subtracted from the width; however, these are to be added, not subtracted.

296. **The correct answers are downward and negative.** First, get $10x - 5x^2 + 7 = y$ in standard form, $y = ax^2 + bx + c$, by rearranging the terms: $y = -5x^2 + 10x + 7$. Since a determines the direction that the parabola opens, identify a as –5. This tells us that the parabola will open downward and have steep walls and a narrow bell shape.

To identify the y-intercept of a quadratic from its equation, move it into standard form, $y = ax^2 + bx + c$, and identify c.

$$y + 7 = 10x - 5x^2$$
$$\underline{-7 -7}$$
$$y = -5x^2 + 10x - 7$$

Since $c = -7$, the y-intercept of the quadratic is –7.

297. **The correct answer is (–3,–4).** The vertex of a quadratic is the maximum or minimum point in the parabola. The minimum value of this function occurs at the point (–3,–4).

298. **The correct answer is choice c.** Notice that this question is not simply asking you to evaluate the function for a *numerical* value of x, but instead you will substitute in the algebraic term, $3v$, for both x's in the function and then simplify the function by performing the required operations:

$$f(x) = 2x^2 + 5x$$
$$f(3v) = 2(3v)^2 + 5(3v)$$
$$f(3v) = 2(3v)(3v) + 5(3v)$$
$$f(3v) = 2(9v^2) + 15v$$
$$f(3v) = 18v^2 + 15v$$

Choice **a** is incorrect because you accidentally multiplied the 2 by the 3 in the term $2(3v)^2$ before applying the exponent. The order of operations dictates that you must apply exponents before performing multiplication.

Choice **b** is incorrect because you accidentally multiplied the 2 by the 3 in the term $2(3v)^2$ before applying the exponent. You also incorrectly combined the two terms together with addition, which you cannot do since $36v^2$ and $15v$ are not like terms.

Choice **d** is incorrect because you incorrectly combined the two terms together with addition, which you cannot do since $18v^2$ and $5v$ are not like terms.

299. **The correct answer is choice b.** Table A represents a function since the only input values that repeat are 3 and both of these input values have the same output value, 10. Table B also represents a function. Even though all of the output values are the same at $q = 5$, every input value is unique and has one and only one output value. Table C is not a function because there are repeated input values that all have different output values, so this violates the function definition that every input must have one and only one output.

300. **The correct answer is choice a.** Similarly, calculate $f(2)$:

$f(x) = 4x + 10$

$f(2) = 4 \cdot 2 + 10$

$f(2) = 18$

Therefore, $f(5) - f(2)$ is calculated as $30 - 18 = 12$.

Choice **b** is incorrect because you interpreted $f(5) - f(2)$ as meaning $5 - 2$, but you must evaluate 5 and 2 in the $f(x)$ function.

Choice **c** is incorrect because 30 is the value of $f(5)$, but you forgot to subtract the value of $f(2)$.

Choice **d** is incorrect because 18 is the value of $f(2)$, but you forgot to subtract this from the value of $f(5)$.

301. **The correct answer is choice a.** This type of problem once again requires you to determine the equation for which the points in the table are true. You can graph these functions, but that requires too much work and can be hard. The easiest method that you can use is simple trial and error with each answer. This method is illustrated next.

$y = x^2 - 1$

If you substitute $(2,3)$ into this equation, you will find:

$3 = (2)^2 - 1$

$3 = 4 - 1$

$3 = 3$ **This is true.**

If you substitute $(3,8)$ into the equation, then the result is:

$8 = (3)^2 - 1$

$8 = 9 - 1$

$8 = 8$ **This is true.**

If you substitute $(4,15)$ into the equation, you will find:

$15 = (4)^2 - 1$

$15 = 16 - 1$

$15 = 15$ **This is true.**

Finally, substitute $(5,24)$ into the formula and:

$24 = (5)^2 - 1$

$24 = 25 - 1$

$24 = 25$ **This is true.**

Thus, this is the correct answer.

302. **The correct answer is choice c.** Factor the left side as
$(3x + 1)(3x + 2)$. Setting each of these factors equal to zero yields
$x = -\frac{1}{3}, -\frac{2}{3}$. So, this equation has two negative real solutions.
Choice **a** is incorrect because one of the solutions of this equation
is zero and the second is the positive real number 2.
Choice **b** is incorrect because the left side factors as $(5x + 1)^2$, so
this equation has a repeated negative real solution, not two differ-
ent negative real solutions.
Choice **d** is incorrect because this equation has two complex conju-
gate solutions.

303. **The correct answer is choice c.** Solve the equation as follows:

$$2\sqrt{1 - x} - 10 = 0$$
$$2\sqrt{1 - x} = 10$$
$$\sqrt{1 - x} = 5$$
$$1 - x = 25$$
$$x = -24$$

Since this value satisfies the original equation, it is the one and only
solution.
Choice **a** is incorrect. When solving a linear equation $ax + b = c$,
subtract b from both sides; do not add it.
Choice **b** is incorrect because $\sqrt{a - b} \neq \sqrt{a} - \sqrt{b}$.
Choice **d** is incorrect because you dropped the radical sign in the
middle of the solution. Once you take 10 to the right side and
divide by 2, the equation should be $\sqrt{1 - x} = 5$, not $1 - x = 5$.

304. **The correct answer is choice d.** Since it is stated that $g(x)$ is a
function that represents the total amount of money Greg earns for
selling x refrigerators in a single day, the input, x, cannot represent
money, hours, or even just commission income. The input, x, must
represent the number of refrigerators sold in a single day, and the
output, $g(x)$, will represent the total amount of money he will earn
that day.
Choice **a** is incorrect because $g(13)$ would represent Greg's total
income on a day he sold 13 refrigerators.
Choice **b** is incorrect because $g(8)$ would represent Greg's total
income on a day he sold 8 refrigerators.
Choice **c** is incorrect because $g(40)$ would represent Greg's total
income on a day he sold 40 refrigerators.

305. **The correct answer is 10 feet.** In order to determine the height that the beanbag was thrown from, look at the y-intercept. The y-intercept is at $(0,10)$, which means that at 0 seconds the beanbag was 10 feet off the ground. Therefore it was thrown from a height of 10 feet.

306. **The correct answers are 0.3 and 12.** The vertex can be used to identify when the beanbag was at its peak and how high it was from the ground. The vertex is approximately $(0.3,12)$, which indicates that after 0.3 seconds, the beanbag reached a maximum height of 12 feet.

307. **The correct answer is 1.1 seconds.** The x-intercept, $(1.1,0)$, tells us that after 1.1 seconds the beanbag was 0 feet from the ground, so this was the point at which it hit the ground.

7

Geometry Foundations

Geometry is the study of shapes and spatial relationships. The geometry skills that will be assessed on the GED test will also benefit you in the real world: working with perimeter, area, volume, and surface area of shapes. Whether you're buying carpeting for your home, grass seed for your lawn, or fencing to protect your dog, having some geometry under your belt will be useful in your everyday life. Get ready to test your proficiency working with the following concepts:

- Perimeter

- Using the Pythagorean theorem with right triangles

- Circumference

- Area

- Surface area of prisms

- Volume of prisms

- Applying scale factors to geometric shapes

308. The perimeter of a square is 24 inches. What is its area?
   a. 144 in.^{2}
   b. 576 in.^{2}
   c. 16 in.^{2}
   d. 36 in.^{2}

309. Find the area of the shaded region in this figure. Remember that the formula for the area of a circle is $A = \pi r^2$.

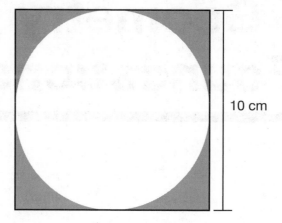

10 cm

   a. 100 cm^2
   b. 78.5 cm^2
   c. 21.5 cm^2
   d. 178.5 cm^2

310. The surface area of a cube is given by the expression $6s^2$, where s is the length of an edge. If a cube has a surface area of 54 square centimeters, what is the length of its edges?
   a. 3 cm
   b. 6 cm
   c. 9 cm
   d. 81 cm

311. If the edge of a cube is 10 cm and the edge of a second cube is 8 cm, what is the difference in the surface areas of the two cubes?
   a. 216 cm^2
   b. 384 cm^2
   c. 788 cm^2
   d. 600 cm^2

312. Find the area of the following shape.

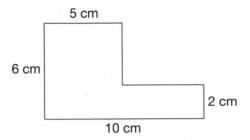

a. 60 cm²
b. 32 cm²
c. 50 cm²
d. 40 cm²

313. The distance between Hamden and Milford is 1.75 cm on a map. In real life, Hamden is 105 km from Milford. On the same map, Cheshire is 2 cm from Mystic. How far is Cheshire from Mystic in real life?
a. 210 km
b. 3.5 km
c. 120 km
d. 107 km

314. The following figure is a regular octagon. What is the perimeter of the figure? _____

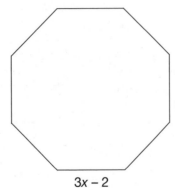

$3x - 2$

315. Jim works for a carpeting company. His next job is to recarpet an office space. According to the diagram, how many square feet of carpet does he need to complete this job?

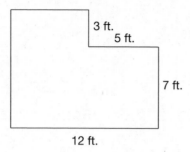

3 ft.

5 ft.

7 ft.

12 ft.

a. 44 ft.²
b. 105 ft.²
c. 120 ft.²
d. 144 ft.²

316. The perimeter of a rectangle is 64. The length of one of the sides of the rectangle is 8. Find the lengths of the other three sides.
a. 10, 23, 23
b. 8, 22, 22
c. 8, 24, 24
d. 12, 22, 22

317. If the following figure is increased by a scale factor of 4, what will the perimeter of the new shape be? _____

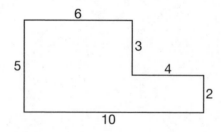

6

3

5   4

2

10

318. Keyonna is reading the plans for an apartment she is decorating. She uses her ruler to see that in the drawing the kitchen measures 2 inches wide by 3.5 inches long. What is the area in square feet of the kitchen? _____

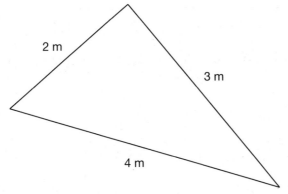

319.

In meters, what is the perimeter of the given triangle?
a. 3 m
b. 6 m
c. 7 m
d. 9 m

320. A right triangle has legs of length 7 and 4. To the nearest tenth, what is the length of its hypotenuse?
a. 3.3
b. 5.7
c. 8.1
d. 11.0

321. The following figure is a rectangle with a half circle attached.

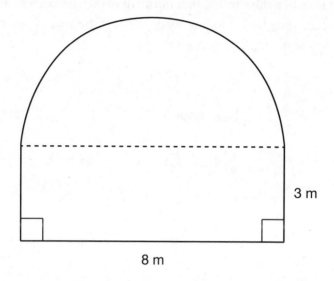

3 m

8 m

Given the indicated dimensions, what is the area of the region in terms of π?

a. $14 + 4\pi$ meters

b. $14 + 16\pi$ meters

c. $24 + 8\pi$ meters

d. $24 + 16\pi$ meters

322. The following figure represents a composite part to be manufac-
tured by fusing together two solid cubes.

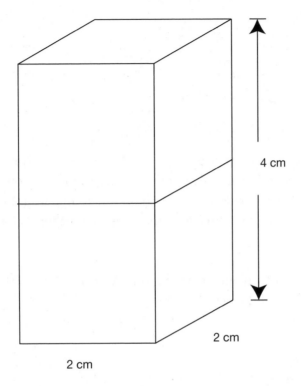

4 cm

2 cm

2 cm

If the cubes used are identical, what is the volume of the resulting
part?
a. 4 cm³
b. 8 cm³
c. 16 cm³
d. 40 cm³

323. The diameter of a circle is 10 meters. In meters, which of the following is the circumference of this circle?
a. 5π
b. 10π
c. 25π
d. 100π

324. What is the missing side length of the following triangle?

_____

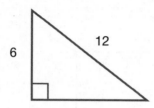

325. Reggie bikes to work every day, going 10 miles north from his house on Robert Street and then going 8 miles east on Dodd Road. He can bike at 18 mph. How much time would Reggie save if he used the bike path that cuts straight through the park?

_____

326. One side of a regular octagon has a length of 4 cm. What is the perimeter of the octagon?
   a. 8 cm
   b. 24 cm
   c. 36 cm
   d. 32 cm

327. Gary makes a diorama out of a shoebox. The shoebox has a width of 6 inches, a length of 12 inches, and a height of 5 inches. What is the volume of Gary's diorama?
   a. 72 in.^{3}
   b. 216 in.^{3}
   c. 360 in.^{3}
   d. 432 in.^{3}

328. Roy draws an equilateral triangle and every side is 6 centimeters long. What is the perimeter of Roy's triangle?
   a. 6 cm
   b. 12 cm
   c. 18 cm
   d. 216 cm

329. The following figure shows a right triangle with sides of length 14, 48, and x. What is the value of x?

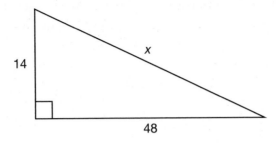

a. 50
b. 55
c. 62
d. 672

330. Rusty needs to order enough wood to fence in the community garden grounds, which are in the shape of a regular pentagon— a pentagon with five congruent sides. If the fencing costs $12.50 per foot, how much will it cost to fence in the community garden?

_____

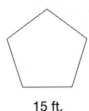

15 ft.

331. Emma works for a catering company, and her boss asked her to line the edges of the serving dishes with a bright red ribbon for Valentine's Day. How much ribbon will Emma need to wrap around a circular pan that can hold a pizza that has a 14-inch diameter? Round your answer to the nearest inch. _____

332. If the circumference of a circle is 62.8 mm, what is its diameter?

_____

333. Melanie is making two ladybug costumes for her 2-year-old twin daughters. She bought a black, rectangular piece of felt fabric that is 1.5 feet × 2.5 feet. She wants to cut as many circles as possible with a 4-inch diameter. How many circles can Melanie cut out of the black felt fabric?

a. 3.75
b. 7
c. 43
d. 44

334. If the perimeter of the square is 24 cm, what is x?

$(x - 3)$

a. 3 cm
b. 6 cm
c. 9 cm
d. 12 cm

335. Tani is carpeting a square room for a client. She charges three times as much for the area of the carpet as she does for installing the baseboard. The price, p, is multiplied by the area of the room and the amount of baseboard used. Which expression indicates how much Tani will charge for this square room?

c

a. $c^2p + 12cp$
b. $3c^2p + 4cp$
c. $12c^3p^2$
d. $c^2 + 4c$

336. The board members of an apartment complex decide that they want to designate 200 ft.² of the common space to make a rectangular picnic area with a grill and some tables. If one of the board members suggests that the length of this space be 25 feet long, how wide would the area be? _____

337. A total of 32 feet of fencing is needed to enclose a square chicken coop. What is the area of this chicken coop? _____

338. Calculate the area of the triangle:

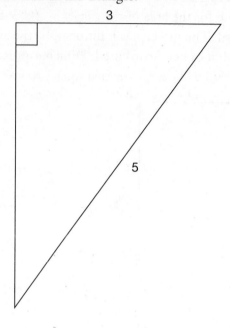

    a. 12 units²

    b. 7.5 units²

    c. 6 units²

    d. 15 units²

339. What is the area of the following rectangle?

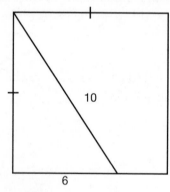

    a. 30

    b. 60

    c. 64

    d. 150

340. A semicircle is inscribed in a rectangle as shown in the following diagram. What is the area in square units of the shaded region?

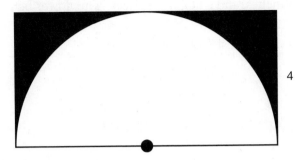

4

a. 16π
b. 24π
c. $32 - 16\pi$
d. $32 - 8\pi$

341. A circle has a diameter of 200 inches. If the diameter is increased by 25%, by how many inches will the circumference of the circle increase?

a. 25π
b. 40π
c. 50π
d. 250π

342. If the perimeter of the triangle is 36, what is its area?

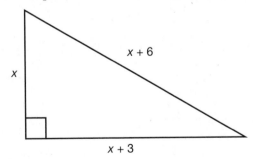

a. 9
b. 12
c. 54
d. 108

343. Tracey is making a flower box to hang underneath a window in the front of her house. Tracey wants the flower box to have a base that is 9 inches by 6 inches, and a height of 2 feet. The flower box will not have a top. If the wood costs $1.50 per square foot, how much is this project going to cost Tracey?

a. $8.63
b. $252
c. $8.07
d. $234

344. As part of her art installation, Tabatha will put a full circle of glitter in the center of the museum gallery floor. Her glitter circle will be 9 feet in diameter. If each vial of glitter contains enough to cover 12 square feet, how many vials of glitter must Tabatha purchase? (*Note:* Only full vials of glitter are available for purchase, so your answer must not contain a decimal.) _____

345. Brenda has hired a landscaper to turn her backyard into a unique hangout spot. She would like to have a circular deck with a diameter of 10 feet built in the middle of her rectangular backyard, which is 25 feet by 18 feet. Since the yard is bare soil right now, she is going to purchase sod to go around the deck. If the contractor charges $1.20 per square foot of sod installed, how much will the purchase of the sod cost Brenda?

a. $78.50
b. $445.80
c. $450.00
d. $528.50

346. Find the missing value in the given trapezoid if its area is 45 cm².

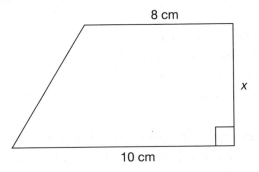

8 cm

x

10 cm

Note: The formula for the area of a trapezoid is provided on the Formula Reference Sheet. It is $A = \frac{1}{2}h(b_1 + b_2)$. _____

347. A local arboretum wants to build a bridge across a pond. At the moment, the walkway around the pond has the measurements shown in the following figure:

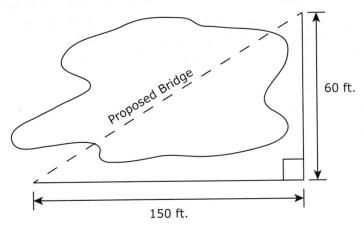

Proposed Bridge

60 ft.

150 ft.

Which expression gives the length of the bridge?

a. $\sqrt{150^2 - 60^2}$

b. $150^2 + 60^2$

c. $60 + 150$

d. $\sqrt{150^2 + 60^2}$

348. The diameter of sphere A is 4 centimeters. What is the volume (in cubic centimeters) of a sphere whose radius is triple that of sphere A?
  a. $2,304\pi$ cubic centimeters
  b. $6,912\pi$ cubic centimeters
  c. 576π cubic centimeters
  d. 288π cubic centimeters

349. The area of a rectangle is 88 square feet, and its perimeter is 38 feet. Which of the following equations can be used to determine the length of the sides of the rectangle?
  a. $2x + 2(88 - x) = 38$
  b. $2x + \frac{176}{x} = 38$
  c. $x + \frac{88}{x} = 38$
  d. $2x + \frac{x}{44} = 38$

350. An airline allows rectangular carry-on bags whose maximum diagonal measurement of a side of the bag is 20 inches and maximum height of that side is 9 inches. Which of the following gives the dimensions of the maximum length (in inches) allowed?
  a. $20^2 - 9^2$
  b. $\sqrt{9^2 + 20^2}$
  c. $\sqrt{20^2 - 9^2}$
  d. $20^2 + 9^2$

351. What is the volume of the following right prism in terms of x?

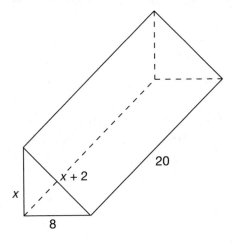

Note: Volume of a right prism is $V = Bh$. _____

352. Find the volume of the cylinder shown here.

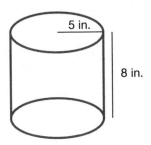

Note: Volume of a cylinder is $V = \pi r^2 h$. _____

353. An empty cylindrical can has a height of 4 inches and a base with a radius of 1.5 inches. Melanie fills the can with water. What is the volume of the water Melanie pours into the can?

a. 5.5π cubic inches
b. 6π cubic inches
c. 6.5π cubic inches
d. 9π cubic inches

354. The length of a side of square A is twice as long as a side of square B. How much larger is the area of square A?
a. 4 times larger
b. 2 times larger
c. 8 times larger
d. 0.5 times larger

355. A 9-foot-long ladder is placed against the side of a building such that the top of the ladder reaches a window that is 6 feet above the ground. To the nearest 10th of a foot, what is the distance from the bottom of the ladder to the building?
a. 1.7 feet
b. 2.4 feet
c. 6.7 feet
d. 10.8 feet

356.

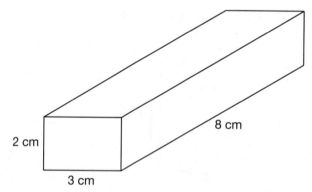

What is the volume of this figure?
a. 6 cm³
b. 24 cm³
c. 48 cm³
d. 108 cm³

357. The surface area of a sphere is 36π cubic meters. To the nearest meter, what is the diameter of this sphere?
a. 3 m
b. 6 m
c. 12 m
d. 24 m

358. Find the radius of the cone if the volume is 148 cm³ and the height is 7 cm. Round your answer to the nearest tenth of a centimeter.

_____

359. The Great Pyramid at Giza has a square base with a side length of approximately 750 feet. The pyramid is approximately 450 feet tall. How many cubic feet of stone were used to build this pyramid?

_____

360. These two boxes have the same volume ($V = l \times w \times h$). Find the length of the missing side on box B.

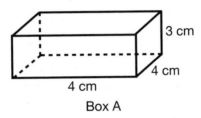

Box A

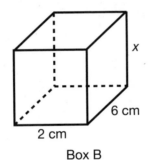

Box B

    a. 3

    b. 4

    c. 5

    d. 6

361. Find the area of the shaded portion in the following figure.

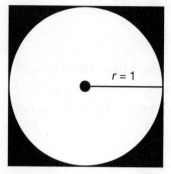

$r = 1$

a. $\pi + 1$
b. $\pi - 1$
c. $2 - \pi$
d. $4 - \pi$

362. What is the value of x in the following figure?

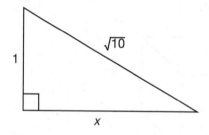

$\sqrt{10}$

1

x

a. 3
b. 4
c. 5
d. 9

363. What is the area of the shaded figure?

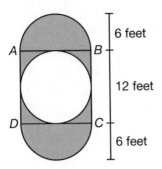

a. 144 square feet – 12π square feet
b. 12π square feet – 144 square feet
c. 144 square feet
d. 6 square feet – 24π square feet + 12π square feet

364. Find the surface area of the following prism:

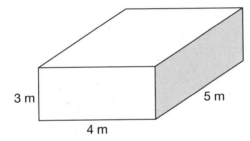

a. 94 m²
b. 68 m²
c. 60 m²
d. 64 m²

365. Find the surface area of a cylinder that has a diameter of 12 cm and a height of 20 cm. _____

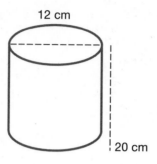

12 cm

20 cm

366. Find the surface area of the given cone.

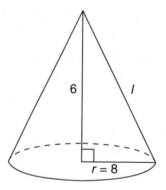

6

l

$r = 8$

    a. 64π units^2
    b. 144π units^2
    c. 80π units^2
    d. 128π units^2

367. The formula for the circumference of a circle of radius r is $2\pi r$. The following shape is formed by attaching a square, whose sides measure 6 meters, to a semicircle whose diameter is 6 meters. What is the perimeter of the shape? Use the approximation 3.14 for π and round your answer to the nearest tenth.

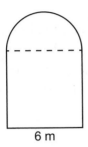

6 m

    a. 27.4 m
    b. 33.4 m
    c. 36.8 m
    d. 39.4 m

368. Ana is making cookies. She rolls out the dough to a rectangle that is 18 inches by 12 inches. Her circular cookie cutter has a circumference of 9.42 inches. Assuming she reuses the dough scraps, approximately how many cookies can Ana cut out of the dough?
    a. 30 cookies
    b. 14 cookies
    c. 12 cookies
    d. 10 cookies

369. Eva and Carr meet at a corner. Eva turns 90° left and walks 5 paces; Carr continues straight and walks 6 paces. If a line segment connected them, it would measure
    a. $\sqrt{22}$ paces.
    b. $\sqrt{25}$ paces.
    c. $\sqrt{36}$ paces.
    d. $\sqrt{61}$ paces.

370. The Great Pyramid at Giza has a square base with a side length of approximately 750 feet. The pyramid is approximately 450 feet tall and the slant height is approximately 783 feet. What is the surface area of the sides of the Great Pyramid that are exposed to sunlight and oxygen (excluding the base)? _____

371. John James High School is making a basketball court in its outdoor playground. The following is a picture of the key to be made in front of each net. If Beto needs to tape around the perimeter of the key so that he can paint it accurately, how many feet of tape does he need?

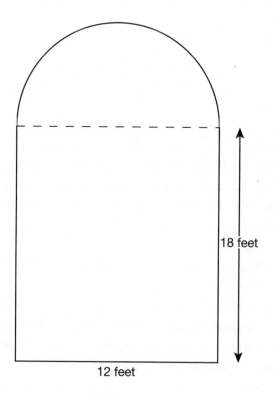

18 feet

12 feet

Round your answer to the nearest whole foot. _____

372. What is the area of a circle with a circumference of 16π?

    a. 64π

    b. 8π

    c. 32π

    d. 256π

373. The diameter of a circle is 16 cm. What is the area (in square centimeters) of a circle whose radius is triple the radius of such a circle?

    a. 24π

    b. 48π

    c. 96π

    d. 576π

374. A cylinder has height 30 inches and base diameter D inches. What is the volume of a cylinder (measured in cubic inches) obtained by doubling the height and taking one-third of the diameter?

    a. $\frac{5}{9}\pi D^2$

    b. $\frac{4}{3}\pi(\frac{D}{6})^3$

    c. $\frac{5}{3}\pi D^2$

    d. $10\pi D$

375. An audience member at a water fountain display wants to determine how high the highest geyser shoots. She measures its shadow as 13 feet, and she knows that a small geyser of x feet casts a shadow of y feet onto the ground. What is the height (in feet) of the highest geyser?

    a. $\frac{x}{13y}$

    b. $\frac{13+x}{y}$

    c. $13xy$

    d. $\frac{13x}{y}$

376. Find the surface area of the right circular cone that has a radius of 10 mm and a slant length of 25 mm. _____

377. How many cubic feet of purified water will be needed to fill a koi pond that is 18 feet long, 3 feet wide, and 3 feet deep, if the waterline of the pond will sit 4 inches from the top edge?
a. 144 ft.²
b. 151.2 ft.²
c. 162 ft.²
d. 183.6 ft.²

378. If the area of a square is 52 in.², what is its perimeter to the nearest tenth of an inch? _____

379. A homeowner designed a deck on which he will place a hot tub. The hot tub is to be enclosed by a 12′ × 12′ cabana. If the homeowner wants two feet of deck surrounding the cabana, what is the distance from point A to point B in the diagram?

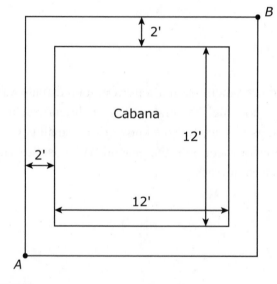

a. $\sqrt{14^2 + 14^2}$
b. $\sqrt{14^2 + 16^2}$
c. $14^2 + 14^2$
d. $\sqrt{16^2 + 16^2}$

380. The surface area of a cube is 150 square centimeters. What is its volume (in cubic centimeters)?
   a. 750 cubic centimeters
   b. 125 cubic centimeters
   c. 5 cubic centimeters
   d. 25 cubic centimeters

381. The area of a rectangle is 96 square inches, and its perimeter is 40 inches. Which of the following systems can be used to determine the dimensions of the rectangle?

   a. $\begin{cases} 2x + 2y = 96 \\ xy = 40 \end{cases}$

   b. $\begin{cases} x + y = 20 \\ \frac{1}{2}xy = 96 \end{cases}$

   c. $\begin{cases} 2x + 2y = 40 \\ \frac{1}{2}xy = 96 \end{cases}$

   d. $\begin{cases} x + y = 20 \\ xy = 96 \end{cases}$

Answers and Explanations

308. **The correct answer is choice d.** All sides of a square are the same length. The perimeter is the distance around the outside of a figure. You can divide the perimeter of a square by 4 to determine the length of a side: 24 ÷ 4 = 6. Therefore, the length of a side of the square is 6 inches. To find the area, multiply the length by the width. In a square, the length and the width are the same. In this case, they are both 6 inches, and 6 × 6 = 36. The area of the square is 36 square inches.

Choice **a** is incorrect because you divided the perimeter of 24 by 2, to find the side length of 12, but you should have divided it by 4, since there are 4 sides in a square.

Choice **b** is incorrect because you used 24 as the side length in the $A = s^2$ formula, but 24 was the perimeter and not the side length, so you needed to find the side length first.

Choice **c** is incorrect because you determined that the side length was 4, but when dividing 24 by 4 you should have found out that the side length was 6.

309. **The correct answer is choice c.** To find the area of the shaded region, subtract the area of the circle from the area of the square. Notice that the radius of the circle is half the length of one side of the square. Therefore, the radius is 5 cm.

Area of square: 10 × 10 = 100 cm²
Area of circle: 3.14 × 5² = 3.14 × 25 = 78.5 cm²
Area of shaded region: square – circle
100 – 78.5 = 21.5 cm²

The area of the shaded region is 21.5 cm².

Choice **a** is incorrect because 100 cm² is the area of the entire square, but you forgot to subtract the area of the circle in order to determine the area of just the shaded corners.

Choice **b** is incorrect because 78.5 cm² is the area of the circle, but you forgot to subtract this from the area of the square in order to determine the area of just the shaded corners.

Choice **d** is incorrect because you should have subtracted the area of the circle from the area of the square, not added it, in order to determine the area of just the shaded corners.

310. **The correct answer is choice a.** The surface area of the cube is the product of 6 and a number squared. So, you can write the equation $6s^2 = 54$ and solve it for s:

$$6s^2 = 54$$
$$s^2 = 9$$

Because $s^2 = 9$, each edge measures 3 cm.

Choice **b** is incorrect because 6 is the coefficient in the surface area formula, but it does not represent the edge length.

Choice **c** is incorrect because when you arrived at $s^2 = 9$, you forgot to take the square root of both sides.

Choice **d** is incorrect because when you arrived at $s^2 = 9$, instead of taking the square root of both sides, you squared both sides.

311. **The correct answer is choice a.** Since each edge of a cube has the same length, the area of each face is s^2. There are six faces on every cube, so the surface area of a cube is $6s^2$.

The surface area of the first cube is:

$$6(10^2) = 6(100) = 600$$

The surface area of the second cube is:

$$6(8^2) = 6(64) = 384$$

The difference between the two surface areas is:

$$600 - 384 = 216$$

Choice **b** is incorrect because 384 cm² is the surface area of the second cube, but not the difference between the two cubes.

Choice **c** is incorrect because 788 cm² is larger than the individual surface areas of the two cubes, so there is no way it could be the difference.

Choice **d** is incorrect because 600 cm² is the surface area of the first cube, but not the difference between the two cubes.

312. **The correct answer is choice d.** Find the lengths of the two missing sides.

The horizontal missing side can be found by subtracting the 5 cm side from the 10 cm side. Therefore, the horizontal missing side is 5 cm.

The vertical missing side can be found by subtracting the 2 cm side from the 6 cm side across from it. Therefore, the vertical missing side is 4 cm.

The following drawing shows all of the sides.

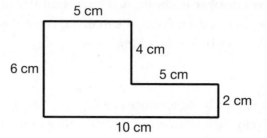

The shape can be broken into two rectangles (two possible ways are shown).

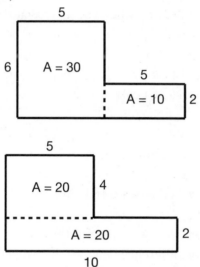

In the first figure, the area of the larger rectangle is 30 cm², and the smaller rectangle is 10 cm². The total area is 40 cm² (30 + 10).

In the second figure, the area of the top rectangle is 20 cm², and the area of the bottom rectangle is 20 cm². The total area is 40 cm² (20 + 20).

Choice **a** is incorrect because 60 cm² would be the area of the complete rectangular figure if it were not missing the top right corner. You multiplied 6 by 10 but then forgot to account for the concave corner.

Choice **b** is incorrect because you added all the sides together and found the perimeter and did not calculate the area.

Choice **c** is incorrect because you multiplied the bottom length of 10 and the top length of 5, but did not account for the width of the figure, or for the top right corner that is missing.

313. The correct answer is choice c.

Method 1:

Set up a proportion comparing the distance in real life and the distance on the map:

$$\frac{\text{map Hamden to Milford}}{\text{real-life Hamden to Milford}} = \frac{\text{map Cheshire to Mystic}}{\text{real-life Cheshire to Mystic}}$$

$$\frac{1.75x}{105} = \frac{2}{x}$$

$$1.75x = 210$$

$$x = 120$$

The distance between Cheshire and Mystic is 120 km.

Method 2:

Determine the number of kilometers represented by 1 cm on the map: $105 \div 1.75 = 60$.

Each centimeter on the map is 60 km in real life.

The distance from Cheshire to Mystic on the map is 2 cm. Since $2 \times 60 = 120$, the distance from Cheshire to Mystic in real life is 120 km.

Choice **a** is incorrect because 210 is cross product found while solving the proportion, but you forgot to divide both sides by 175.

Choice **b** is incorrect because you set up your proportion incorrectly. You must put like terms in the same part of each ratio, but instead of inputting 2 as the distance from Hamden to Milford, you input it as the distance from Cheshire to Mystic.

Choice **d** is incorrect because you added 105 km and 2 cm, which are values that represent real life and map distance, respectively.

314. The correct answer is 24*x* – 16. The perimeter of a figure is the distance around it. For a regular octagon (whose sides all have equal lengths), the perimeter can be found by multiplying the length of one side times the total number of sides. According to the diagram, the length of each side of the octagon is $3x - 2$, so the perimeter is $8(3x - 2)$.

Be sure to distribute the 8 to both terms inside the parentheses so as not to arrive at $24x - 2$, which is incorrect.

When the 8 is distributed correctly,

$$P = 8(3x) - 8(2)$$
$$= 24x - 16$$

315. **The correct answer is choice b.** There are two ways to solve this problem. The first is to divide the room into two rectangles, calculate the area of each, and add the areas together. There are two ways to divide the room into two rectangles:

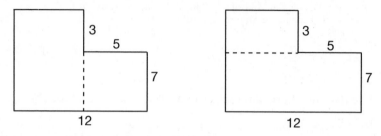

In the first diagram, one rectangle has a length of 7 feet and a width of 5 feet, and the other has a length of 10 feet, resulting from 7 + 3, and a width of 7, resulting from 12 – 5. To find the area of the room, set up the following equation: (5 ft.)(7 ft.) + (10 ft.)(7 ft.). This yields 105 ft.² as the area of the room.

In the second diagram, one rectangle has a length of 7 feet and a width of 12 feet, and the other has a width of 7 feet, resulting from 12 – 5, and a length of 3 feet. To find the area of the room, set up the following equation: (12 ft.)(7 ft.) + (7 ft.)(3 ft.). This also yields 105 ft.² as the area of the room.

The second method is to calculate the area of the big rectangle— (12 ft.)(10 ft.)—and subtract the area of the part of the room that is missing—(5 ft.)(3 ft.). This equation reads 120 ft.² – 15 ft.² = 105 square feet.

Choice **a** is incorrect because you calculated the perimeter and not the area of the room.

Choice **c** is incorrect because you calculated the area of the complete rectangle, but you forgot to subtract the 3-by-5-foot corner that is missing from the top right.

Choice **d** is incorrect because 144 ft.² could be the area if both the width and the length of this room were 12 feet, and if it wasn't missing the 3-by-5-foot corner from the top right.

316. **The correct answer is choice c.** The perimeter of a figure is the distance around the figure. Since the opposite sides of a rectangle are equal, and one side of this rectangle has a length of 8, another side also has a length of 8. The set of numbers whose sum is 64 when added to 8 is represented by 8 + 8 + 24 + 24 = 64.

Choices **a** and **d** are incorrect because a rectangle must have two pairs of equal side lengths, but these dimensions each display only one pair of equal side lengths.

Choice **b** is incorrect because these side lengths result in a perimeter of 60 and not the given perimeter of 64.

317. **The correct answer is 120 units.** The perimeter of the original compound shape will be the sum of all of its sides: 5 + 6 + 3 + 4 + 2 + 10 = 30 units. When a figure is increased by a scale factor, all of its individual sides are multiplied by that factor. Therefore, all of the sides here would be multiplied by 4 in order to determine the side lengths and associated perimeter of the new figure: 5(4) + 6(4) + 3(4) + 4(4) + 2(4) + 10(4) = 120 units. A shortcut to finding the new perimeter after using a scale factor of 4 is to simply multiply the original perimeter by 4: 30 units × 4 = 120 units.

318. **The correct answer is 112 ft.².** Since each inch in the drawing represents 4 feet in real life, multiply the scaled dimensions by 4 to get their real-life dimensions:

2 inch width × 4 feet = 8 feet wide

3.5 inch length × 4 feet = 14 feet long

Since the kitchen is 8 feet wide by 14 feet long, multiply these two dimensions to get the area: 8 feet × 14 feet = 112 square feet.

319. **The correct answer is choice d.** The perimeter is the sum of all side lengths: 2 + 4 + 3 = 9.

Choice **a** is incorrect. The area of the triangle is 3 square meters, not the perimeter.

Choices **b** and **c** are incorrect. The lengths of all sides must be added to find the perimeter, not just two of them.

320. **The correct answer is choice c.** Using the Pythagorean theorem, $7^2 + 4^2 = c^2$ where c is the length of the hypotenuse. Solving for c, $c^2 = 65$ and $c = \sqrt{65} \approx 8.1$.

Choice **a** is incorrect. The Pythagorean theorem requires that all of the terms be squared, not just the length of the hypotenuse. Choice **b** is incorrect. When using the Pythagorean theorem $a^2 + b^2 = c^2$, a and b represent the lengths of the legs. Choice **d** is incorrect. While the Pythagorean theorem does contain a sum, the terms are also squared.

321. **The correct answer is choice c.** The area of the rectangular region is $8 \times 3 = 24$ square meters, while the area of the half circle is $\frac{1}{2}\pi r^2 = \frac{1}{2}\pi(\frac{8}{2})^2 = \frac{1}{2}\pi(16) = 8\pi$.

Choice **a** is incorrect. This is the perimeter of the region. Choice **b** is incorrect. This would be the perimeter of the region if the radius was 8 meters (this is the diameter) and if it was a full circle instead of a half circle. Choice **d** is incorrect. The area of the half circle is half of the usual area formula πr^2. This is the area if the full circle was used.

322. **The correct answer is choice c.** The volume of one of the cubes is $2 \times 2 \times 2 = 8$ cm^3. Since the part consists of two cubes, the final volume is double this, or 16 cm^3.

Choice **a** is incorrect. This is the area of one face of one of the cubes. Choice **b** is incorrect. This is the volume of only one of the cubes used to make the part. Choice **d** is incorrect. This is the surface area of the final part.

323. **The correct answer is choice b.** The radius of the circle is 5, and the circumference is $2 \times \pi \times$ (radius), or 10π. This can also be found simply by multiplying the diameter and π.

Choice **a** is incorrect. The radius of the circle is 5 and must be doubled in order to find the circumference. Choice **c** is incorrect. This is the area of the circle, which is found by squaring the radius and multiplying by π. Choice **d** is incorrect. The diameter does not need to be squared in order to find the circumference.

324. The correct answer is $a = 10.4$. Substitute the given values into the Pythagorean theorem. Note that the missing side is *not* the hypotenuse, so let's use a as the missing side length.

$$a^2 + 6^2 = 12^2$$
$$a^2 + 36 = 144$$
$$a^2 + 36 = 144$$
$$\underline{\quad -36 \quad -36 \quad}$$
$$\sqrt{a^2} = \sqrt{108}$$
$$a \approx 10.4$$

325. The correct answer is 17 minutes. First, let's find how long it takes Reggie to get to work using his normal route. He bikes 10 miles north and then 8 miles east for a total of 18 miles. If he bikes at 18 mph, how long does it take him to get to work? Use the distance = rate × time formula and solve for time:

$$d = rt$$
$$18 = 18t$$
$$\frac{18}{18} = \frac{18t}{18}$$
$$1 \text{ hour} = t$$

It takes Reggie 1 hour to get to work using his normal route. To find how long it would take him to get to work using the bike path, we need to find its distance using the Pythagorean theorem.

$$a^2 + b^2 = c^2$$
$$10^2 + 8^2 = c^2$$
$$100 + 64 = c^2$$
$$164 = c^2$$
$$\sqrt{164} = \sqrt{c^2}$$
$$12.8 = c$$

The distance from Reggie's house to work using the bike path is 12.8 miles. Use the distance formula to find how long it will take Reggie to get to work using the bike path:

$$d = rt$$
$$12.8 = 18t$$
$$0.71 \text{ hours} = t$$

To find how much faster Reggie would get to work using the bike path versus his normal route, subtract the two times.

$$0.71 \text{ hours} = (0.71)(60) = 42.6 \approx 43 \text{ minutes}$$
$$60 \text{ minutes} - 43 \text{ minutes} = 17 \text{ minutes}$$

So, if Reggie uses the bike path to get to work, he will get to work 17 minutes faster than by taking his normal route.

326. **The correct answer is choice d.** A regular octagon has eight sides, all of which are the same length. The perimeter is the length around the outside of a figure. If all eight sides of 4 cm each are added up, the perimeter is 32 cm.

Choice **a** is incorrect because 8 cm would be the length of only two sides, but an octagon has eight sides.

Choice **b** is incorrect because 24 cm would be the perimeter of a regular triangle with a side length of 8 cm, but you needed to find the perimeter of an octagon.

Choice **c** is incorrect because you multiplied the side length of 4 cm by 9 to get 36 cm, but an octagon has eight sides, not nine.

327. **The correct answer is choice c.** The formula for the volume of a rectangular prism is: (length)(width)(height). Substitute the length, width, and height of the diorama into the formula:

Volume of a rectangular prism = (12 in.)(6 in.)(5 in.)

Volume = 360 in.^{3}

Choice **a** is incorrect because it is the area of one side of the diorama. You are looking for the volume of the diorama.

Choice **b** is incorrect because the formula for the volume of a cube is s^3, where s is the length of one side of the cube. You would arrive at 216 in.^{3} if you took the width of the diorama, 6 inches, and used the formula for the volume of a cube. The diorama, however, is a rectangular prism, not a cube.

Choice **d** is incorrect because the width of the shoebox is 6 inches, but the height is only 5 inches. You would arrive at 432 in.^{3} if you multiplied $6 \times 12 \times 6$.

328. **The correct answer is choice c.** The formula for perimeter is given in the question; add the lengths of each side of the triangle to find the perimeter of the triangle. Roy draws an equilateral triangle, which is a triangle with three sides that are all the same length. All three sides are 6 cm long:

6 cm + 6 cm + 6 cm = 18 cm

Choice **a** is incorrect because 6 cm is the length of only one side of the triangle.

Choice **b** is incorrect because 12 cm is the length of only two sides of the triangle.

Choice **d** is incorrect because you could take the length of one side of the triangle and multiply it by 3 ($6 \times 3 = 18$) to find the perimeter of the triangle, but you cannot take the length of one side of the triangle and raise it to the third power ($6^3 = 216$).

329. **The correct answer is choice a.** Use the Pythagorean theorem to determine the length of the third side of the triangle.

$14^2 + 48^2 = c^2$, where c is the hypotenuse

$14^2 = 196$ and $48^2 = 2{,}304$

So, $c = \sqrt{196 + 2{,}304}$ or $\sqrt{2{,}500}$, which equals 50.

330. **The correct answer is $937.50.** Since the garden is in the shape of a regular pentagon, the side lengths are equivalent. Therefore, the perimeter is

$p = 5(15 \text{ ft.})$

$= 75 \text{ ft.}$

The cost of the fencing is $12.50 per foot, so the total cost to fence in the community garden is

$C = (75 \text{ ft.})(\$12.50)$

$= \$937.50$

331. **The correct answer is 44 inches.** Use the circumference formula with diameter since that is the dimension provided:

Circumference $= \pi d$

$C = \pi(14)$

$C = 43.96$, which rounds to 44 inches of ribbon.

332. **The correct answer is 20 mm.** Remember, the equation for circumference is $C = 2\pi r$ or $C = \pi d$. Since we need to find the diameter to solve this problem, let's use $C = \pi d$.

$C = \pi d$

$62.8 \text{ mm} = (3.14)d$

$20 \text{ mm} = d$

333. **The correct answer is choice c.** First, find the area of the piece of black fabric. $A_{fabric} = lw = (2.5 \text{ ft.})(1.5 \text{ ft.}) = 3.75 \text{ ft.}^2$. Since the fabric is in feet, the unit length of the diameter of the circles needs to be converted from inches to feet. To find the area of a circle, the radius—not the diameter—is needed. If the diameter is 4 inches, the radius is 2 inches. Therefore, the radius is 2 inches out of a possible 12 inches in a foot: $\frac{2}{12} = \frac{1}{6}$. Next, find the area of one circle Melanie wants to cut. Area $= \pi r^2 = \pi(\frac{1}{6} \text{ ft.})^2 = 0.087 \text{ ft.}^2$. The number of circles Melanie will be able to cut can be found by dividing the area of the fabric, 3.75 ft.^{2}, by the area of each circle, 0.087 ft.^{2}: $\frac{3.75}{0.087} = 43.1 = 43$ circles.

Choice **a** is incorrect because it gives the area of the felt fabric.

Choice **b** is incorrect because it results if the radius is not squared when solving for the area of each circle.

Choice **d** is incorrect because it results from mistakenly rounding up to 44 but there is only enough fabric for one-tenth more than 43 circles.

334. **The correct answer is choice c.** The perimeter of a square is the sum of all four sides. Thus, $P_{square} = (x-3) + (x-3) + (x-3) + (x-3) = 24$ cm. After gathering like terms, the equation is $4x - 12 = 24$ cm.

$$4x - 12 = 24$$
$$\underline{+12 \ +12}$$
$$4x = 36$$
$$\frac{4x}{4} = \frac{36}{4}$$
$$x = 9 \text{ cm}$$

Choice **a** is incorrect because it results if the four −3 terms are incorrectly grouped to be +12 when solving for x.

Choice **b** is incorrect because it is the length of each side.

Choice **d** is incorrect because you might have made an error in addition when adding 24 + 12, thinking the sum was 48.

335. **The correct answer is choice b.** The area of the room is $c \times c = c^2$. The perimeter of the room is $c + c + c + c = 4c$. It costs three times as much to carpet the room, $3p$, as it does to install the baseboard, p. So the total expression that illustrates the cost of the project is $3p(c^2) + p(4c)$. Rearranging the equation so that the coefficients come first and the exponents are written in descending order, an equivalent expression is $3c^2p + 4cp$.

Choice **a** is incorrect because it multiplies the wrong terms by $3p$ and p.

Choice **c** is incorrect because it multiplies the costs together instead of adding them.

Choice **d** is incorrect because it does not factor in the cost of each installation.

336. **The correct answer is 8 feet.** Since this is going to be a rectangular picnic area, use the area formula for a rectangle, plug in the given dimensions for the area and the length, and work backward to see what the width would be:

Area = length × width

$200 = 25 \times w$

$\frac{200}{25} = w$, so the width would be 8 feet. That would be a pretty narrow picnic area!

337. **The correct answer is 64 square feet.** The first piece of information is the perimeter of the coop, since 32 feet *encloses* the coop. Once we find the side length, we can use that in the area formula for a square. Since we are working with a square coop, use the perimeter formula for a square, plug in the perimeter of 32 feet, and work backward to obtain the side length:

Perimeter = $4s$

$32 = 4s$

So, $s = 8$. Now, plug this into the area formula:

Area = s^2

Area = $8^2 = 64$ ft.^{2}

338. **The correct answer is choice c.** In order to calculate the area of any triangle, you must have the dimensions of its base and height, which are perpendicular to one another. You have two perpendicular sides in this triangle, but do not know both of their dimensions. You are given the hypotenuse of 5 and one of the legs of 3, so use this information to solve for the missing leg in the Pythagorean theorem:

$a^2 + b^2 = c^2$

$(3)^2 + b^2 = (5)^2$

$9 + b^2 = 25$

$b^2 = 16$

$b = 4$

Now that you know that the two perpendicular sides measure 3 units and 4 units, you can put them into the area formula for triangles:

$A = \frac{1}{2}bh = \frac{1}{2}(4)(3) = 6$ units^2

Choice **a** is incorrect because you identified the missing leg and found the product of the base times the height, but then you forgot to take half of that product. Remember that the formula for the area of a triangle is $A = \frac{1}{2}$(base)(height).

Choice **b** is incorrect because you used the leg and the hypotenuse in the area formula, but you should have used the two legs of the triangle, since the base and height are always perpendicular to each other.

Choice **d** is incorrect because you multiplied the two given sides together without regard for the fact that in the formula $A = \frac{1}{2}$(base) (height) the base and height must be perpendicular to one another, and you neglected to take half of this product.

339. The correct answer is choice c. There is a right triangle with a hypotenuse of 10 and a leg of 6. Using the Pythagorean theorem, this makes the height of the rectangle 8. The diagram shows that the height and width of the rectangle are equal, so it is a square. 8^2 is 64.

Choice **a** is incorrect because you likely multiplied the given values and divided by 2, but neither side represents a side of the given rectangle, and you do not divide by 2 when finding the area of a rectangle.

Choice **b** is incorrect because you simply multiplied 10×6, which will not give you the area of the rectangle, as neither represents a side of the rectangle.

Choice **d** represents a calculation error.

340. The correct answer is choice d. The area of the shaded region is the difference between the area of the rectangle and the area of the semicircle.

Use the fact that the semicircle is inscribed inside the rectangle to label other parts of the figure. Since the semicircle is inscribed in the rectangle, the radius of the circle is the same as the height of the rectangle.

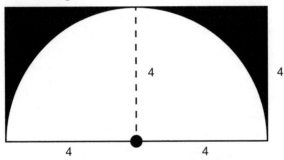

The area of the rectangle is:

$A = bh$

$A = (8)(4) = 32$

$A = \frac{1}{2}\pi r^2$

$A = \frac{1}{2}\pi(4)^2$

The area of the semicircle is:

$A = \frac{1}{2}\pi(16)$

$A = 8\pi$

The difference between the two areas is $32 - 8\pi$.

341. **The correct answer is choice c.** Use the original diameter to calculate the circumference of the original circle:

$C = 2\pi r$ or πd

$C = \pi(200)$

$C = 200\pi$

Next, increase the diameter by 25%: new diameter = 200 +
200(0.25) = 200 + 50 = 250.

Use the increased diameter to calculate the circumference of the
new circle:

$C = \pi d = \pi(250) = 250\pi$

Calculate the increase in the circumference by subtracting:

$250\pi - 200\pi = 50\pi$

342. **The correct answer is choice c.** Since the perimeter is given, set
up an addition problem to find x:

$(x + 3) + (x + 6) + x = 36$

$3x + 9 = 36$

$3x = 27$

$x = 9$

Since $x = 9$, the measurements of the triangle are known:

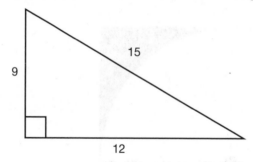

The area of the triangle is $\frac{1}{2}bh = \frac{1}{2}(12)(9) = 54$.

343. **The correct answer is choice c.** The equation for surface area of a rectangular prism is needed to solve this problem: $SA = ph + 2B$, where p is the perimeter of the base and B is the area of the base. Because the price is calculated per square foot, the 6 inches and 9 inches need to be converted to feet, 0.5 feet and 0.75 feet respectively, before substituting the values into the formula. Once all the values are plugged into the equation, it reads $SA = (0.5 + 0.5 + 0.75 + 0.75)(2) + 2(0.5)(0.75)$. Before going further, the formula needs to be modified for this problem because the flower box will not have a top. Only one of the bases should be calculated. So, the surface area of the flower box is $(2.5)(2) + (0.5)(0.75)$, which yields 5.375 ft.². When multiplied by $1.50, we find the total cost of the project to be $8.07 (rounded up from $8.0625).

Choice **a** is incorrect because it does not take into account that the flower box does not have a top.

Choice **b** reflects neither the conversion of units nor the fact that only one base needs to be calculated.

Choice **d** does not take into account that inches need to be converted to feet.

344. **The correct answer is 6.** First, determine that the radius of the circle will be 4.5 feet, and use this to calculate the area of the circle:

$$A = \pi r^2$$
$$A = \pi(4.5)^2$$
$$A = 63.6 \text{ ft.}^2$$

Since each vial of glitter will cover only 12 square feet, divide 63.6 by 12 to determine how many vials are needed: 5.3. Although numerically 5.3 is closer to 5 than it is to 6, Tabatha will need 6 vials of glitter since 5 vials will not be enough to make a fully glittered 9-foot circle.

345. The correct answer is choice b. To find the cost of the sod, we need to find the area of the rectangular yard and then subtract the area of the circular deck.

$$A_{yard} = \text{length} \times \text{width}$$
$$= (25 \text{ ft.})(18 \text{ ft.})$$
$$= 450 \text{ ft.}^2$$
$$A_{deck} = \pi r^2$$
$$= (3.14)(5 \text{ ft.})^2$$
$$= 78.5 \text{ ft.}^2$$
$$A_{sod} = 450 \text{ ft.}^2 - 78.5 \text{ ft.}^2$$
$$= 371.5 \text{ ft.}^2$$

Last, multiply the sodded area by the cost per square foot to get the total cost of installation of sod.

$$\text{Cost} = (371.5)(\$1.20)$$
$$= \$445.80$$

Choice **a** is incorrect because this is the area of just the deck, without the square footage of the yard taken into account. Also, you did not incorporate the cost of the sod.

Choice **c** is incorrect because this is the area of the entire yard, without the deck removed from the square footage. Also, you did not take into account the cost of the sod.

Choice **d** is incorrect because this is the combined area of the yard and the deck, but instead of adding them, you should have subtracted them and then multiplied that difference by the cost of sod per square foot.

346. The correct answer is 5 cm. Since we are given the area of both the bases, we can put these into the trapezoid area formula and work backward to find the height, x:

$$A = \tfrac{1}{2}h(b_1 + b_2)$$
$$45 = \tfrac{1}{2}h(10 + 8)$$
$$45 = \tfrac{1}{2}h(18)$$
$$45 = h(9), \text{ so } h = 5 \text{ cm}$$

347. **The correct answer is choice d.** Apply the Pythagorean theorem using the lengths of the legs as 60 and 150. The length of the bridge is $\sqrt{60^2 + 150^2}$ feet.

Choice **a** is incorrect because you should add the squares of the legs, not subtract them.

Choice **b** is incorrect because you forgot the square root.

Choice **c** is incorrect because you cannot simplify a square root of a sum in this manner: $\sqrt{a^2 + b^2} \neq a + b$. Or you might have incorrectly assumed that the longest side is equal to the sum of the shorter sides.

348. **The correct answer is choice d.** The radius of the given sphere must be 2 cm. So the radius of the desired sphere is 3(2 cm) = 6 cm. So its volume is $\frac{4}{3}\pi(6 \text{ cm})^3 = 288\pi$ cubic centimeters.

Choice **a** is incorrect because you used the diameter instead of the radius when computing the volume.

Choice **b** is incorrect because you used the diameter instead of the radius when computing the volume, and you forgot to include the $\frac{1}{3}$ in the volume formula for a sphere.

Choice **c** is incorrect because you used the diameter instead of the radius and used the surface area formula instead of the volume formula.

349. **The correct answer is choice b.** Let x be the length and y be the width. Since the area is 88 square feet, we have $xy = 88$, which implies that $y = \frac{88}{x}$. Since the perimeter is 38 feet, we know that $2x + 2y = 38$, so that substituting in $y = \frac{88}{x}$ yields the equation $2x + 2\left(\frac{88}{x}\right) = 38$, which is equivalent to $2x + \frac{176}{x} = 38$.

Choice **a** is incorrect because the two sides x and y are not related by $x + y = 88$; you misused the information regarding the area.

Choice **c** is incorrect because you forgot to multiply the width and length by 2 in the perimeter formula.

Choice **d** is incorrect because using the area formula, the two sides of the rectangle are x and $\frac{88}{x}$, not x and $\frac{x}{88}$.

350. **The correct answer is choice c.** Begin by drawing a diagram:

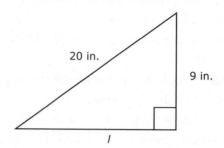

Using the Pythagorean theorem, we obtain $9^2 + l^2 = 20^2$. Solving for l then yields $l^2 = 20^2 - 9^2$, so that $l = \sqrt{20^2 - 9^2}$.

Choice **a** is incorrect because you forgot the square root.

Choice **b** is incorrect because 20 is the hypotenuse, not a leg, of the triangle.

Choice **d** is incorrect because you forgot the square root, and 20 is the hypotenuse, not a leg, of the right triangle.

351. **The correct answer is $80x$ cm^3.** Starting with the formula for the volume of a right prism $V = Bh$, we see that we have to first find the area of the base, B. Since this is a triangular prism, find B by using the formula for the area of a triangle:

Area of triangle base $= \frac{1}{2}bh$

The base and height must be perpendicular to each other, so we will use the triangle's side lengths of 8 and x in the formula:

Area of triangle base $= \frac{1}{2}(8$ cm$)(x$ cm$) = 4x$ cm^2

Use $4x$ cm^2 as the B in the volume formula and use 20 cm as the height of the prism, h:

$V = Bh$

$V = (4x$ cm$^2)(20$ cm$)$

$V = 80x$ cm^3

352. **The correct answer is 628 in.^{3}.** The formula for the volume of a cylinder is $V = \pi r^2 h$. From the diagram, we can see that the height of the cylinder is 8 inches and the radius is 5 inches. Substitute these values into the formula to solve for the volume.

$V = \pi r^2 h$

$\quad = \pi(5$ in.$)^2(8$ in.$)$

$\quad = \pi(25$ in.$^2)(8$ in.$)$

$\quad = (3.14)(200$ in.$^3)$

$\quad = 628$ in.^{3}

353. **The correct answer is choice d.** Use the formula $V = \pi r^2 h$, where r is the radius of the base and h is the height of the cylinder: $\pi(1.5^2)4 = \pi \times 2.25 \times 4$, which equals 9π cubic inches. Choices **a** and **c** are incorrect due to a calculator error. Choice **b** is incorrect because you forgot to square the radius of 1.5 inches before multiplying it by the height of 4 inches.

354. **The correct answer is choice a.**

Method 1:

Choose a few examples of the given situation and analyze the results.

Example: If square A has sides of length 10, square B will have sides of length 5.

Then, the area of A is 100, and the area of B is 25.

The area of square A is 4 times the area of square B.

Example: If square A has sides of length 6, square B will have sides of length 3.

Then, the area of A is 36, and the area of B is 9.

The area of square A is 4 times the area of square B.

If you continue to try other situations, the results will be the same. The area of the larger square is always 4 times the area of the smaller square.

Method 2:

The situation can be analyzed algebraically.

Length of side of square B = x

Length of side of square A = $2x$

Area of square B = x^2

Area of square A = $(2x)^2 = 4x^2$

$4x^2$ is 4 times x^2.

Choice **b** is incorrect because when finding the area of a square you will be multiplying *two* sides together that are *each* twice as long, so the area will not simply be increased by a factor of 2, but by a factor of 2 *times* 2, which is 4.

Choice **c** is incorrect because this is the factor by which the *volume* would increase when the side lengths of a square are all doubled, but the area will increase by a factor of 2 just twice, so it will be 4 times as large.

Choice **d** is incorrect because square A has longer sides than square B, so there is no way that its area could be smaller than that of square B.

355. **The correct answer is choice c.** Using the Pythagorean theorem, the hypotenuse of the right triangle formed by the ladder and the building is 9, while the length of one leg is 6. This yields the equation $6^2 + b^2 = 9^2$ or $b^2 = 81 - 36 = 45$. Therefore, $b = \sqrt{45} \approx$ 6.7 feet.

Choice **a** is incorrect. The terms in the Pythagorean theorem are squared.

Choice **b** is incorrect. Applying the Pythagorean theorem to this problem yields the equation $6^2 + b^2 = 9^2$. The exponent of 2 indicates to multiply the term by itself twice, not multiply by 2.

Choice **d** is incorrect. The length of the ladder represents the hypotenuse, or c, in the Pythagorean theorem.

356. **The correct answer is choice c.** The area of the base is $2 \times 3 =$ 6 square centimeters. Multiplying this by the height of 8 cm gives us the volume in cubic centimeters: $6 \times 8 = 48$ cm³.

Choice **a** is incorrect. This is the area of one of the smaller faces.

Choice **b** is incorrect. This is the area of one of the larger faces.

Choice **d** is incorrect. This is the surface area of the given shape.

357. **The correct answer is choice b.** Using the surface area formula:

$$36\pi = 4\pi r^2$$
$$9 = r^2$$
$$r = 3$$

Since the radius is 3, the diameter is $3 \times 2 = 6$ m.

Choice **a** is incorrect. This is the radius of the sphere. The diameter is twice as large as the radius.

Choice **c** is incorrect. When solving the equation $36\pi = 4\pi r^2$, divide, do not multiply, both sides by 4. Additionally, the diameter will be two times as large as the radius.

Choice **d** is incorrect. When solving the equation $36\pi = 4\pi r^2$, divide (do not multiply) both sides by 4.

358. **The correct answer is 4.5 cm.** The formula for the volume of a cone is $V = \frac{1}{3}r^2h$. We have values for V and h:

$V = 148$ cm^3

$h = 7$ cm

We then substitute these values into the formula, giving us the following equation with r as the only unknown:

$V = \frac{1}{3}r^2h$

$148 \text{ cm}^3 \quad = \frac{1}{3}\pi r^2(7 \text{ cm})$

$(3)148 \text{ cm}^3 = (3)\frac{1}{3}\pi r^2(7 \text{ cm})$

$444 \text{ cm}^3 \quad = (3.14)(7 \text{ cm})r^2$

$\frac{444 \text{ cm}^3}{(3.14)(7 \text{ cm})} = \frac{(3.14)(7 \text{ cm})r^2}{(3.14)(7 \text{ cm})}$

$20.2 \text{ cm}^2 \quad = r^2$

$\sqrt{20.2 \text{ cm}^2} = \sqrt{r^2}$

$\qquad 4.5 \text{ cm} = r$ (rounded to 10th)

Rounded to the nearest tenth of a centimeter, the volume is 4.5 cm.

359. **The correct answer is 84,375,000 ft.^{3}.** The volume formula for the area of a right pyramid requires that we first find the area of the rectangular base of the pyramid. Since the Great Pyramid at Giza has a square base with a side length of approximately 750, multiply 750 by itself to get the area of the square base: $750 \times 750 = 562{,}500$ ft.^{2}. Since we know the height of the Great Pyramid is approximately 450 feet, we are ready to use the volume formula for pyramids:

Volume of a right pyramid $= \frac{1}{3}Bh$

$V = \frac{1}{3}(562{,}500 \text{ ft.}^2)(450 \text{ ft.})$

$V = 84{,}375{,}000$ ft.^{3}

So, the equivalent of more than 84 million 1-foot cubes were used to construct the Great Pyramid!

360. **The correct answer is choice b.** The volume of box A is 48 cm³ (4 × 4 × 3 = 48). The volume of box B must also be 48 cm³, so the three dimensions of box B will multiply to 48. Solve the equation for x:

$$2 \times 6 \times x = 48$$
$$12x = 48$$
$$x = 4$$

Choice **a** is incorrect because box A has a volume of $4 \cdot 4 \cdot 3 = 48$ cm², and box B has a volume of $12x$, so if x were 3, the volume of box B would only be 36 cm².

Choice **c** is incorrect because box A has a volume of $4 \cdot 4 \cdot 3 = 48$ cm², and box B has a volume of $12x$, so if x were 5, the volume of box B would be 60 cm².

Choice **d** is incorrect because box A has a volume of $4 \cdot 4 \cdot 3 = 48$ cm², and box B has a volume of $12x$, so if x were 6, the volume of box B would be 72 cm².

361. **The correct answer is choice d.** The shaded area is the difference between the area of the square and the circle. Because the radius is 1, a side of the square is 2. The area of the square is $s^2 = 2^2 = 4$, and the area of the circle is $\pi r2 = \pi(1)2 = \pi$. Therefore, the answer is $4 - \pi$.

362. **The correct answer is choice a.** The Pythagorean theorem states that the square of the length of the hypotenuse of a right triangle is equal to the sum of the squares of the other two sides, so we know that the following equation applies: $1^2 + x^2 = 10$, so $1 + x^2 = 10$, so $x^2 = 10 - 1 = 9$, so $x = 3$.

363. **The correct answer is choice c.** This question is much simpler than it seems. The half circles that cap square $ABCD$ form the same area as the circular void in the center. Find the area of square $ABCD$, and that is your answer. 12 feet × 12 feet = 144 square feet.

364. **The correct answer is choice a.** Start with the formula for the surface area of a rectangular prism: $ph + 2B$.

First calculate B, the area of the base, by multiplying the length by the width: $(3\text{ m})(4\text{ m}) = 12\text{ m}^2$. $B = 12\text{ m}^2$. Next, calculate p = perimeter of base. Perimeter $= 2l + 2w = 2(4\text{ m}) + 2(3\text{ m}) = 14\text{ m}$. The height of the prism is 5 m. Plug these three measurements into the surface area formula:

$$SA = ph + 2B$$
$$SA = (14\text{ m})(5\text{ m}) + 2(12\text{ m}^2)$$
$$SA = 70\text{ m}^2 + 24\text{ m}^2$$
$$SA = 94\text{ m}^2$$

Choice **b** is incorrect because you found the total edge length of this rectangular prism and not the surface area.

Choice **c** is incorrect because you found the volume of this rectangular prism and not the surface area.

Choice **d** is incorrect because you forgot to include the left and right sides, which each have a dimension of 3 m by 5 m, so your answer was 30m² too small.

365. **The correct answer is 979.68 cm².** To substitute values into the formula $SA = 2\pi rh + 2\pi r^2$, we first need to identify the radius. The problem states that the diameter is 12 cm. The radius is half of the diameter, so the radius = 6 cm.

Now, substitute values into the formula for the surface area of a cylinder:

$$SA = 2\pi rh + 2\pi r^2$$
$$= 2\pi(6\text{ cm})(20\text{ cm}) + 2\pi(6\text{ cm})^2$$
$$= 2\pi(120\text{ cm}^2) + 2\pi(36\text{ cm}^2)$$
$$= 2(3.14)(120\text{ cm}^2) + 2(3.14)(36\text{ cm}^2)$$
$$= 753.6\text{ cm}^2 + 226.08\text{ cm}^2$$
$$= 979.68\text{ cm}^2$$

So, the surface area of the cylinder is 979.68 cm².

366. **The correct answer is choice b.** Remember that the surface area for a cone requires the slant height of the cone. Here, the slant height is labeled as l but we can solve for l by using 6 and 8 as the legs in the Pythagorean theorem:

$$a^2 + b^2 = c^2$$
$$6^2 + 8^2 = c^2$$
$$36 + 64 = c^2$$
$$100 = c^2$$
$$c = 10$$

So we now know that the slant height = 10.

Now use the surface area formula, keeping the final answer in terms of π:

Surface area of a cone = $\pi r s + \pi r^2$
$$SA = \pi(8)(10) + \pi(8)^2$$
$$SA = 80\pi \text{ units}^2 + 64\pi \text{ units}^2$$
$$SA = 144\pi \text{ units}^2$$

Choice **a** is incorrect because you found the surface area of the circular bottom part of the cone, but forgot to add this to the conical top part of the cone.

Choice **c** is incorrect because you found the surface area of the conical top part of the cone, but forgot to add this to the circular bottom of the cone.

Choice **d** is incorrect because you found the volume of the cone instead of the surface area.

367. **The correct answer is choice a.** A semicircle is one-half of a circle. The radius of this semicircle is one-half of 6, or 3. So, the perimeter of the semicircular part of the figure is equal to πr, or approximately 3.14×3 meters. The perimeter of the part of the figure made up of the square is 3×6 or 18 meters because only three of the four sides of the square make up the perimeter. Therefore, the perimeter of the entire shape rounded to the nearest tenth is approximately $(3.14 \times 3) + 18$, or about 27.4 meters. Choice **b** is incorrect because all four sides of the square were used to calculate the perimeter of the lower part of the figure. Only three sides of the square make up the perimeter of the figure. Choice **c** is incorrect because it includes the entire circumference of the circle rather than just the circumference of the semicircle. Choice **d** is incorrect because in calculating the circumference of the semicircle, the radius used was 6, rather than 3.

368. **The correct answer is choice a.** The area of the dough is 216 square inches (18×12). To find the area of the cookie cutter, first find the radius. The formula is $C = \pi r^2$. In this instance, $9.42 = 2 \times 3.14 \times r$, or $9.42 = 6.28 \times r$. Divide 9.42 by 6.28 to find r, which is 1.5. The formula for area of a circle is πr^2, in this case, 3.14×1.5^2, or 7.07 square inches. So, the area of the cookie cutter is 7.07 square inches. Divide 216 (the area of the dough) by 7.07 (the area of the cookie cutter), and the result is 30.55, or approximately 30 whole cookies.

369. **The correct answer is choice d.** The corner forms the right angle of this triangle; Eva and Carr walk the distance of each leg, and the question wants to know the hypotenuse. Plug the known measurements into the Pythagorean theorem: $5^2 + 6^2 = c^2$. $25 + 36 = c^2$. $61 = c^2$. Therefore, $\sqrt{61} = c$.

370. **The correct answer is 1,174,500 ft.².** Consider the formula for the surface area of a pyramid: $\frac{1}{2}ps + B$. Since this question asks for the surface area of the sides of the Great Pyramid that are exposed to sunlight and oxygen, excluding the base, that means that we can just ignore the final part of the formula, the "+ B," since that represents the area of the base. Since the side lengths of the square base are 750 feet, we can determine that the perimeter of the base = 4(750 ft.) = 3,000 ft. Put this and the slant height of 783 feet into the shortened formula:

Surface area without the base = $\frac{1}{2}ps$

$SA = \frac{1}{2}(3,000)(783) = 1,174,500$ ft.²

371. **The correct answer is 105 feet.** Since we are being asked to find the perimeter of this shape, we first need to identify that this is a rectangle on the bottom with half of a circle on top. We will not count the top part of the rectangle (where the dotted line is) so this dimension will need to be omitted from the calculations. Label all the sides carefully:

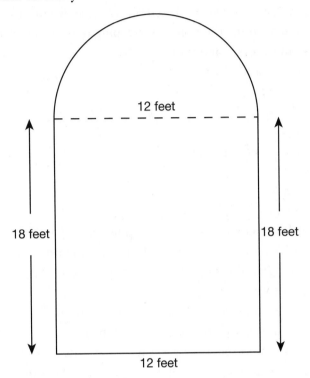

Calculate the partial perimeter of the rectangular base:

Partial perimeter of rectangular base = 18 + 12 + 18 = 48 ft. Since the top of the rectangle has been labeled 12 feet, it should be apparent that the semicircle has a radius of 6 feet. Calculate the circumference of the circular top and then cut it in half to find the circumference of just half the circle:

Circumference of circular top = πr^2

$C = \pi 6^2$

$C = 36\pi \approx 113$ ft.

The semicircular top of the figure has a circumference of $\frac{113}{2}$ = 56.5 ft.

Combine the partial perimeter and half circle circumference to get the complete perimeter of the basketball key: 48 ft. + 56.5 ft. = 104.5 ft. Beto will need about 105 feet of tape to tape off the perimeter of the key.

372. **The correct answer is choice a.** In order to determine the area of a circle, the radius is needed. Since we are not given the diameter or the radius, but instead the circumference, we must work backward to determine the radius. First, plug 6π into the formula for the circumference, and then solve for r:

$C = 2\pi r$, and $C = 16\pi r$

$16\pi = 2\pi r$

$\frac{16\pi}{2\pi} = \frac{2\pi r}{2\pi}$

$8 = r$

Next, plug $r = 8$ into the area formula:

$A = \pi r^2$

$A = \pi(8)^2$

$A = 64\pi$

Choice **b** is incorrect because you plugged 16π into the area formula, solved for r, and then used r in the circumference formula. You were *given* the circumference, so you should have done these steps in reverse.

Choice **c** is incorrect because you simply doubled the circumference, but this is not how the circumference and area are related.

Choice **d** is incorrect because you mistook the 16 in the 16π circumference as being the radius, and used $r = 16$ in the area formula, instead of solving for r by working backward.

373. **The correct answer is choice d.** Since the diameter is 16 cm, the radius is 8 cm. So, the radius of the desired circle is 24 cm. Its area is $\pi(24)^2 = 576\pi$ square centimeters. Choice **a** is incorrect because you forgot to square the radius. Choice **b** is incorrect because it is the circumference, not the area. Choice **c** is incorrect because you computed the circumference using the diameter instead of finding the area.

374. **The correct answer is choice c.** The height of the new cylinder is 2(30 in.) = 60 in. The diameter of the base of the new cylinder is $\frac{D}{3}$ in., so its radius is $\frac{D}{6}$ in. So the volume of the new cylinder is $\pi(\frac{D}{6})^2(60) = \frac{5}{3}\pi D^2$ cubic inches.

Choice **a** is incorrect because you mistakenly used the volume formula for a right circular cone.

Choice **b** is incorrect because you mistakenly used the volume formula for a sphere.

Choice **d** is incorrect because you did not square the radius in the volume formula.

375. **The correct answer is choice d.** Let H be the height of the highest geyser. Set up a proportion: $\frac{H}{13} = \frac{x}{y}$. Solving for H then yields $H = \frac{13x}{y}$.

Choice **a** is incorrect. You should multiply by 13, not divide by it.

Choice **b** is incorrect because, when solving a proportion for a specific quantity, you should cross multiply, not cross add.

Choice **c** is incorrect because you should divide by y, not multiply by it.

376. **1,099 mm².** Since the values for s and r are given in the question, we can substitute them into the formula and solve for surface area:

$$\begin{aligned}
SA &= \pi r s + \pi r^2 \\
&= (3.14)(10 \text{ mm})(25 \text{ mm}) + (3.14)(10 \text{ mm})^2 \\
&= 785 \text{ mm}^2 + 314 \text{ mm}^2 \\
&= 1{,}099 \text{ mm}^2
\end{aligned}$$

377. **The correct answer is choice a.** Since the water will only be filled 4 inches from the top, instead of using 3 feet as the height, subtract 4 inches from 3 feet and use 2 feet, 8 inches as the height. Since there are 12 inches in a foot (and not 10), 2' 8" cannot be represented as 2.8'. Instead, represent it as $2\frac{8}{12}$ feet, which is $2\frac{2}{3}$ or $\frac{8}{3}$ feet. Now, find the volume of a rectangular prism that has a length of 18 feet, a width of 3 feet, and a height of $\frac{8}{3}$ feet:

$$V = l \times w \times h$$
$$V = 18 \times 3 \times \frac{8}{3}$$
$$V = 144 \text{ ft.}^2$$

Choice **b** is incorrect because you translated 2 feet 8 inches to be equivalent to 2.8 feet, but this is not the case; 2 feet 8 inches is actually equal to $2\frac{8}{12}$ feet, which is equivalent to $2.\overline{6}$ feet or $\frac{8}{3}$ feet. Choice **c** is incorrect because you calculated the volume of water needed to fill the pond all the way up to the top. You did not account for the top 4 inches of the pool that will not have water. Choice **d** is incorrect because you used 3.4 feet as the height of the pool instead of subtracting 4 inches from the 3-foot height and using $2.\overline{6}$ feet as the height of the pool.

378. **The correct answer is 28.8 in.** First, you need to find the side length of the square by solving for s using the equation for area:

$$A = s^2$$
$$52 \text{ in.}^2 = s^2$$
$$\sqrt{52 \text{ in}^2} = \sqrt{s^2}$$
$$7.2 \text{ in.} = s$$

Now that we know what s is, we can find the perimeter:

$$P = 4(7.2 \text{ in.})$$
$$= 28.8 \text{ in.}$$

379. **The correct answer is choice d.** The length of the deck is 12 + 2 + 2 = 16, and the width is the same. Using the Pythagorean theorem, the distance from point A to point B is $\sqrt{16^2 + 16^2}$.

Choice **a** is incorrect because you added 2 feet of walkway to the width and to the length of the cabana, but the walkway goes all the way around the cabana; you should have added 4 feet to the width and length of the cabana.

Choice **b** is incorrect because you added 2 feet of walkway to one of sides, but the walkway goes all the way around the cabana; 4 feet should have been added to both the width and the length of the cabana.

Choice **c** is incorrect because you added 2 feet of walkway to the width and to the length of the cabana, but the walkway goes all the way around the cabana; you should have added 4 feet to the width and length of the cabana. And, when applying the Pythagorean theorem, you forgot to include the square root.

380. **The correct answer is choice b.** Let e be the edge length of the cube. The surface area formula is $6e^2 = 150$, so $e^2 = 25$ and $e = 5$. Thus, the volume of the cube is $e^3 = 5^3 = 125$ cubic centimeters.

Choice **a** is incorrect because you used $6e^3$ for the volume formula instead of e^3, where e is an edge of the cube.

Choice **c** is incorrect because this is the length of an edge of the cube, not its volume.

Choice **d** is incorrect because this is the area of a face of the cube, not its volume.

381. **The correct answer is choice d.** Let x denote the width of the rectangle and y the length. The perimeter equation is $2x + 2y = 40$, which is equivalent to $x + y = 20$ (where we have divided both sides by 2), and the area equation is $xy = 96$. So the system in choice **d** is correct.

Choice **a** is incorrect because the right sides of the equations should be interchanged.

Choices **b** and **c** are incorrect because the area equation is incorrect. You used the area formula for a triangle.

8

Interpreting Data in Graphs and Tables

Graphs and tables are used to visually represent information on a wide range of subjects: the rising cost of healthcare, the number and age ranges of people who live in a community, or global warming data. Graphic information is everywhere—television commercials, newspaper reports, and web pages. This chapter assesses how well you can interpret graphs and how successful you are at manipulating data to create tables to accurately represent information. You will be challenged with questions on the following topics:

- Increasing and decreasing intervals
- Positive and negative intervals
- Relative maximums and minimums
- Graph symmetry
- End behavior of functions
- Bar graphs and histograms
- Pie charts
- Dot plots, box plots, and scatter plots

382. Which graph has a *y*-intercept at –5 and increases during the interval $(-\infty, -5) \cup (-5, \infty)$?

a.

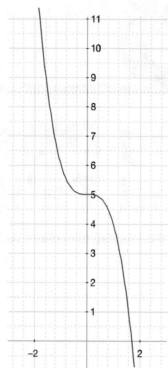

b.

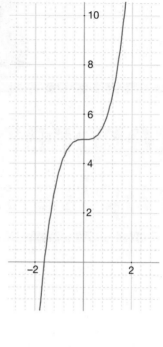

c.

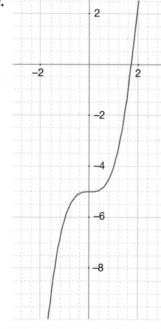

d.

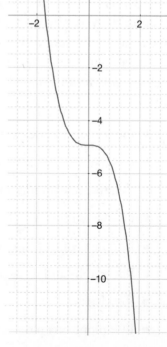

Use the following graph of the function g(x) to answer questions 383 through 386:

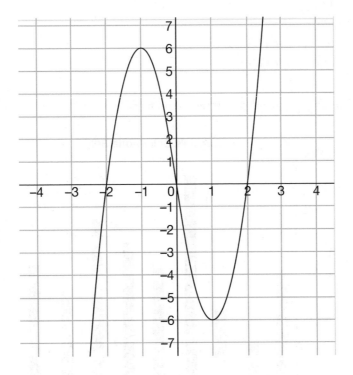

383. On what interval is $g(x)$ decreasing?
   a. $[-2,0]$
   b. $(0,2)$
   c. $(-1,1)$
   d. $(6,-6)$

384. Which of the following functions could be the general form of the equation that represents this function, for real coefficients $a, b, c, d,$ and m?
   a. $y = ax^2 + bx + c$
   b. $y = ax^3 + bx^2 + cx + d$
   c. $y = mx + b$
   d. $y = \frac{1}{x}$

385. What is the best approximation of the relative minimum of $g(x)$?
   a. $(-1,6)$
   b. $(0,0)$
   c. $(1,-6)$
   d. $(-2.4,-7.5)$

386. Over which interval or intervals is g(x) positive?
    a. (0,6)
    b. (−2,0)
    c. (−∞,−1) and (1,+∞)
    d. (−2,0) and (2,+∞)

Use the following bar graph to answer questions 387 and 388:

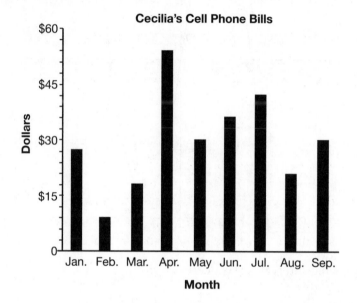

387. Cecilia plots on a bar graph the cost of her cell phone bill for each month from January through September. How much did Cecilia spend on her cell phone in April and May combined?

_____

388. Which answer best approximates the average cost of Cecilia's cell phone bill each month?
    a. $15
    b. $20
    c. $30
    d. $40

Use the following box plot to answer questions 389 and 390:

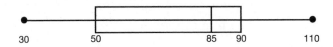

389. This box plot shows the prices of textbooks at a local high school. What range describes the middle 50% of the prices (*p*) of the textbooks?
 a. $30 < p < $85
 b. $50 < p < $85
 c. $50 < p < $90
 d. $85 < p < $90

390. What percentage of books cost between $85 and $90?
 a. 25%
 b. 50%
 c. 75%
 d. It cannot be determined from this graph.

Use the following pie chart to answer questions 391 and 392:

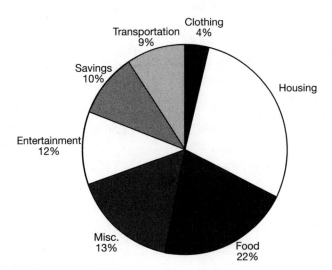

391. The graph shows the Johnson family budget for one month.

In percentage of overall expenses, how much more money is spent on food than on transportation and clothing combined?

_____

392. What percentage does the Johnson family have budgeted for their housing? _____

393. Which of the following statements best describes the relationship between the data points shown on the scatter plot?

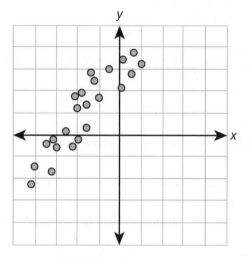

a. There is a positive correlation.
b. There is a negative correlation.
c. There does not appear to be any correlation.
d. It cannot be determined without knowing the values of the data points.

394. Two high-school biology classes hosted a bird-watching day where students kept track of how many different species of birds they observed in a nearby park. The dot plot represents the number of species observed by many of the students.

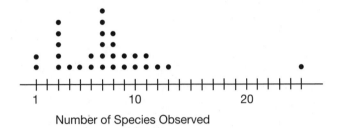

Number of Species Observed

Four of the students have not been included on the plot. The number of species these students observed was:

STUDENT	NUMBER OF SPECIES OBSERVED
Amy	14
Scott	14
Crystal	21
Gilbert	9

Draw as many dots on the graph as is necessary to add these students' observations to the plot.

395. The following histogram represents the data collected through a survey of students at a large commuter college. Each student surveyed provided the one-way distance he or she travels to campus.

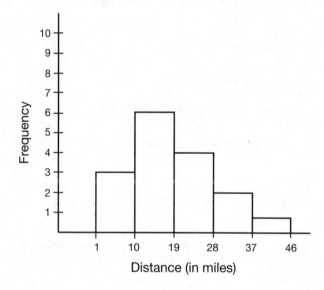

Distance (in miles)

Based on the data, which of the following statements must be true?
a. A total of 46 students were surveyed.
b. There is one student who travels exactly 46 miles to campus, one way.
c. Between 10 and 19 students travel exactly 6 miles to campus, one way.
d. Fewer than 5 students travel less than 10 miles to campus, one way.

396. This figure represents the cumulative number of packages loaded onto trucks in one day at a small warehouse. When the day began, there were already 50 packages loaded.

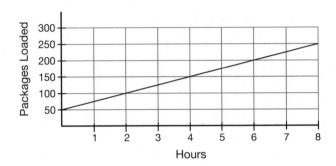

Based on this graph, how many packages were loaded each hour?
a. 25
b. 50
c. 125
d. 250

397. The following chart represents the enrollment in an annual professional training program for several nonconsecutive years. Circle the year for which there was the largest difference between the number of men enrolled and the number of women enrolled in the program.

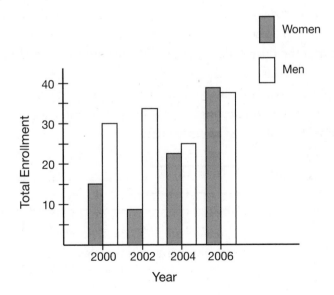

398. The chart represents the number of households in selected cities that have subscribed to a new company's Internet service.

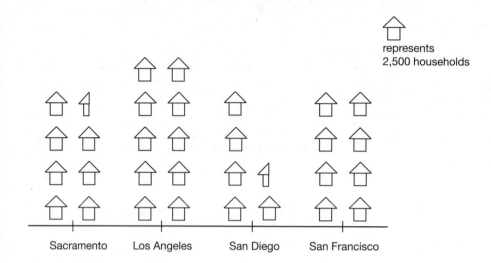

represents
2,500 households

Sacramento   Los Angeles   San Diego   San Francisco

Based on this data, how many households have subscribed to the service in San Diego?
a. 13,750
b. 15,000
c. 18,750
d. 20,000

399. In a study of its employees, a company found that people spent anywhere from 1 to 8 hours a day working on their computers, 50% of the employees spent more than 3 hours on their computers, and the overall distribution of time employees spent on their computers was skewed right. The most frequent response from employees was 4 hours, and the mean time spent on computers was 4.5 hours. Complete the box plot by drawing an additional line so that it matches the given information.

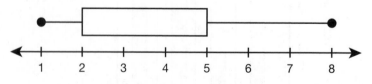

Hours Spent on Computer per Day

400.

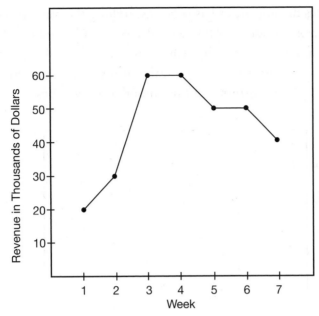

The graph shown here represents the total weekly revenue of a company over several weeks. For which of the following periods has the weekly revenue increased?

a. between weeks 2 and 3
b. between weeks 3 and 4
c. between weeks 4 and 5
d. between weeks 6 and 7

401. A cosmetics manufacturer has been researching the way that people use various products. After several surveys, it has collected the data shown in this scatter plot, which shows the time that participants spent getting ready on a typical morning versus the amount of money the participants spent per month on cosmetics.

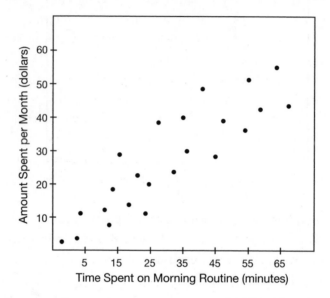

Given this plot, which of the following best describes the relationship between the amount of time spent and the amount of money spent?

a. In general, the longer people spent on their morning beauty routine, the more money they spent per month on beauty products.

b. In general, the longer people spent on their morning beauty routine, the less money they spent per month on beauty products.

c. In general, the amount of time people spent on their morning beauty routine was about the same as the amount of money they spent in dollars on beauty products.

d. In general, there is no clear relationship between the amount of time people spent on their beauty routine and the amount of money they spent per month on beauty products.

402. The bar chart represents the total dollar value of sales for four product versions in July.

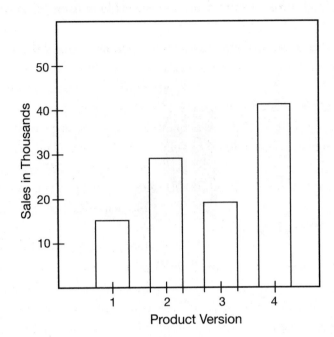

Of the following, which two products had combined sales of more than $50,000 in July?
a. products 1 and 2
b. products 2 and 3
c. products 2 and 4
d. products 1 and 3

403. The bar graph shows the number of families in a small town that have one, two, three, four, or five children per family. What is the mode of the data showing the number of children per family?

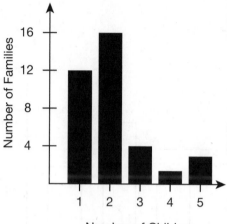

    a. 1
    b. 2
    c. 3
    d. 4

404. This graph shows the number of inches of rain for five towns in Suffolk County during the spring of 2017. What is the average for the five towns?

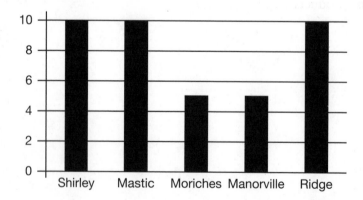

    a. 5
    b. 8
    c. 9
    d. 10

405. What is the median of the data displayed here? _____

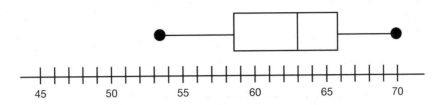

406. The students at Norton School took a survey on the students' favorite kinds of pets. The results are in this table.

TYPE OF PET	NUMBER OF STUDENTS
Dog	30
Cat	25
Fish	2
Bird	3

Ms. Lucian's class wants to put the data into a circle graph (pie chart). How many degrees should the dog sector be?

a. 30°

b. 180°

c. 50°

d. 360°

407. Stephen recorded the number of butterflies that he saw in his backyard for four months and put the information in this table.

MONTH	NUMBER OF BUTTERFLIES
May	28
June	44
July	64
August	56

What was the mean number of butterflies in Stephen's backyard during these four months?

a. 60

b. 54

c. 48

d. 43

408. According to the following chart, how many miles per hour did Rhonda bike?

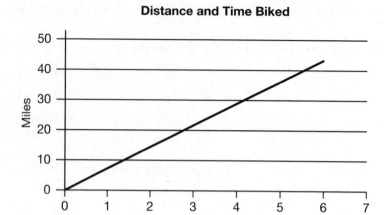

Distance and Time Biked

a. 1 mph
b. 3 mph
c. 7 mph
d. 12 mph

409. Consider the following scenario:

- Rabbits eat a certain type of plant in a neighborhood.
- Initially, these plants are plentiful in the neighborhood and neighbors have many rabbit sightings.
- As time goes on, the plants diminish in number and neighbors see fewer rabbits hopping about.
- As the plants begin to flourish once again, the number of rabbit sightings increases.

Which of the following graphs accurately depicts this scenario?

a.

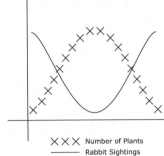

$\times\times\times$ Number of Plants
——— Rabbit Sightings

b.

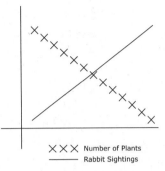

$\times\times\times$ Number of Plants
——— Rabbit Sightings

c.

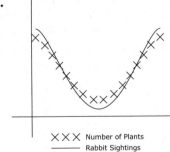

$\times\times\times$ Number of Plants
——— Rabbit Sightings

d.

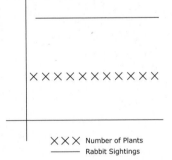

$\times\times\times$ Number of Plants
——— Rabbit Sightings

410. Eight subway passengers are asked to tell a surveyor the number of sneezes they had heard on their trip that day, and then to rate their perception of their exposure to disease on a scale of 1 to 10, where 1 is very low exposure and 10 is very high. The data and the best fit line are shown:

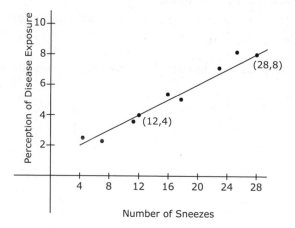

Based on this information, how many sneezes would you expect a passenger to have heard to rate his or her exposure to disease as 10?

a. 40
b. 30
c. 32
d. 36

411. At a recent annual alumni event, the event coordinator asked every twentieth guest who came through the entrance, "How do you prefer to be informed of such events?" She recorded the results in this table:

METHOD OF CONTACT	FREQUENCY
Phone	3
Text message	11
E-mail	19
Newsletter	7

Based on this information, if there are 8,000 alumni total, how many would you expect would prefer being contacted by text message?

a. 3,800
b. 1,400
c. 5,800
d. 2,200

412. Consider the following rectangle:

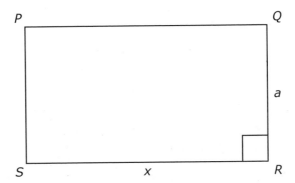

Which expression describes the sum of the lengths of the diagonals $\overline{SQ}$ and $\overline{PR}$ as a function of x?

a. $\sqrt{a^2 + x^2} + \sqrt{a^2 - x^2}$
b. $2a^2 + 2x^2$
c. $2a + 2x$
d. $2\sqrt{a^2 + x^2}$

413. Two different groups, each consisting of 20 people from the audience of a rock concert, are asked to choose their favorite song from a list of four: song A, song B, song C, and song D. The results are as follows:

	SONG A	SONG B	SONG C	SONG D	TOTAL
Group 1	4	4	12	0	20
Group 2	4	6	10	0	20
Total	8	10	22	0	40

Which of the following is an accurate inference regarding the entire audience?

a. Exactly 25% of the audience has song B as their favorite song on this list.

b. One expects that song C would be listed as the favorite by most audience members.

c. More than half of the audience has song C as their favorite on the list.

d. Song C must be the most modern hit based on the number who identified it as their favorite.

414. The following graph shows the frequency of finding certain types of coins on the beach during a scavenger hunt:

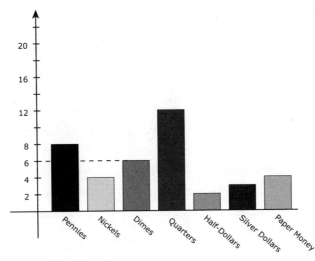

Assuming this is typical of such scavenger hunts, what percentage of the findings obtained in a hunt would you expect to be attributed to either quarters or silver dollars?

a. 40%

b. 30%

c. 10%

d. 20%

415. In a psychology experiment, the following information shows the number of hours slept versus the number of errors made during a 10-minute driving simulation.

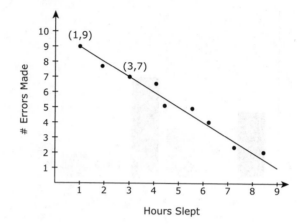

Suppose the best fit line is as shown in the diagram. Use it to predict the number of errors you would expect to make if you had 4 hours of sleep prior to completing the simulation.

a. 10
b. 6
c. 1
d. 5

416. The following chart shows the product ratings for a new smart home electronic device:

RATING (NUMBER OF STARS)	FREQUENCY
0	6
1	3
2	5
3	25
4	12
5	28

What is the median product rating?

a. 3.5 stars
b. 4.0 stars
c. 3.0 stars
d. 3.4875 stars

417. The following line graph shows the monthly electric bills for a customer over the course of a year.

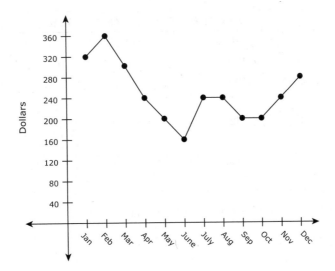

Which two consecutive months show the largest difference in cost?
a. January to February
b. September to October
c. October to November
d. June to July

418. Members of a trading card game group compare their card collections at a weekly meeting. Each member counted the number of rare cards in his or her collection this week. The following are the results:

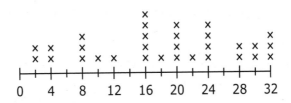

What is the mode number of rare cards in this group?
a. 32
b. 20
c. 16
d. 30

419. The table shows age and average number of text messages sent daily by eight randomly selected mall goers.

AGE	NUMBER OF TEXT MESSAGES
16	120
20	135
28	100
33	83
35	52
41	40
68	18
70	5

The scatter plot with best fit line is shown here:

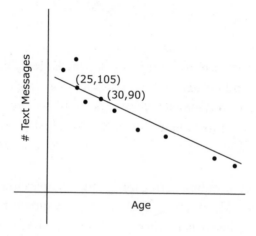

Based on this data, if a person is 50 years old, approximately how many text messages would you expect him or her to send daily?

a. 60

b. 90

c. 30

d. 135

420. Mark the point(s) on the graph to indicate the position(s) of the relative maximum(s). You may indicate more than one spot if necessary:

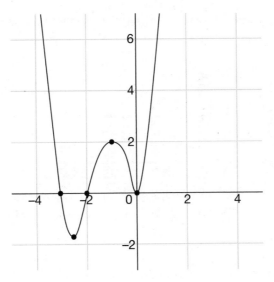

Use the following bar graph to answer questions 421 and 422:

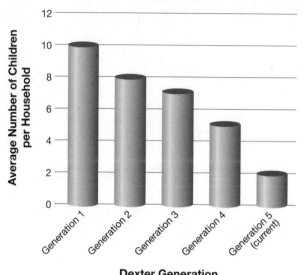

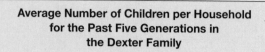

Average Number of Children per Household for the Past Five Generations in the Dexter Family

Dexter Generation

As you can see, the categories—Generation 1, Generation 2, Generation 3, Generation 4, and Generation 5 (current)—are placed along the *x*-axis. The data values—the average number of children per household—are placed along the *y*-axis at equal intervals. At a quick glance, you can see that the average number of children per household has decreased from generation to generation.

421. On average, how many more children did households have in the first generation than the current Dexter generation, according to the graph? _____

422. If the current trend in family size continues for the Dexter family, which is the least likely number of children for the 6th generation?
a. 0
b. 1
c. 2
d. 3

There are five different categories of books at the Everdale Library. Use the following bar graph to answer questions 423 and 424:

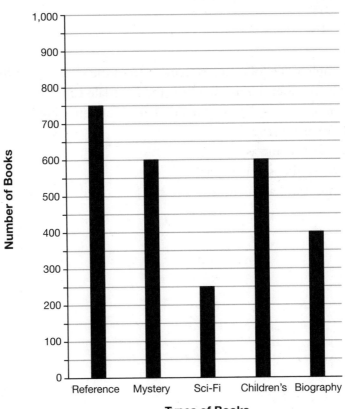

Books at Everdale Library

423. Approximately what percentage of the books at Everdale Library are biographies?

    a. 10%

    b. 15%

    c. 20%

    d. 23%

424. Complete the following statement from the choices that follow: "The number of sci-fi books at the Everdale Library seems to be about _____ the number of reference books."

    a. twice as many as

    b. half as many as

    c. three times as many as

    d. one-third as many as

425. Based on the values in the table, which circle graph accurately represents this data?

U.S. Census Bureau Statistics, 2012

SINGLE MOTHERS WITH CHILDREN UNDER 18	SINGLE FATHERS WITH CHILDREN UNDER 18
10.322 million	1.956 million

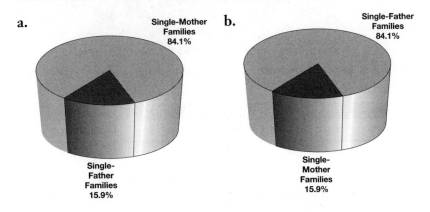

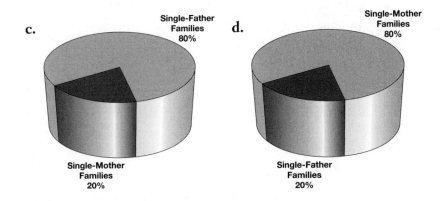

Use the pie chart to answer questions 426 and 427:

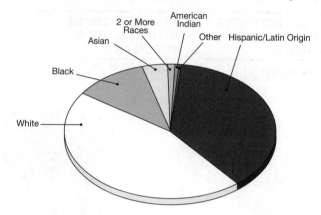

Breakdown of Ethnic Groups in Undisclosed Country

426. If the pie chart represents the breakdown of ethnic groups in a country of 10 million people, what is the best estimate for the number of people of Hispanic/Latin origin in that country?
  a. 5 million
  b. 3 million
  c. 2 million
  d. 1 million

427. Approximately what percentage of the population is white in this country?
  a. 30% to 40%
  b. 40% to 50%
  c. 50% to 60%
  d. 60% to 70%

428. This list of data represents the number of disposable coffee cups the members of a local police force reported using in a given week. Make a dot plot to accurately represent this data: 7, 5, 7, 2, 10, 0, 7, 5, 12, 0, 7, 5, 1, 14, 8

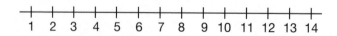

429. Use your dot plot from question 428 to select the correct response from each of the following:

It is *true/false* that the majority of the police force averages at least four disposable coffee cups per week. *1/2/3/4* people on the police force use fewer than two disposable coffee cups per week.

430. The two box plots show the data for daily high temperatures for every day of the fall in Minnesota and Alaska:

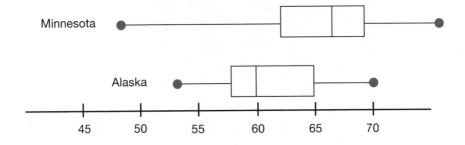

Select one from each underlined section to make the following statement true:

Minnesota/Alaska has a more consistent temperature in the fall, and approximately *25%/50%* of the fall days in Minnesota have a daily high of no more than 62 degrees.

The following histogram displays the age range of students enrolled in Mr. Duvall's Retirement for Rookies class at a local community college. Use this histogram to answer questions 431 to 433:

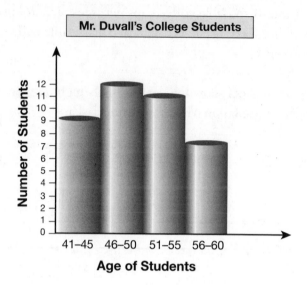

431. How many students are in Mr. Duvall's class?
  a. 50
  b. 44
  c. 39
  d. 35

432. According to the graph, what percentage of students are ages 46 to 50? Round your answer to the nearest tenth. _____

433. According to the graph, what is the percentage difference between the age group with the highest number of students and the age group with the lowest number of students? _____

Use the scatter plot comparing hours studied per week to weeks to pass the GED test to answer questions 434 and 435:

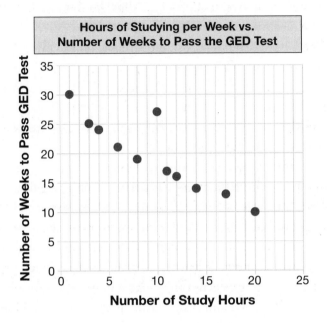

434. If each of the following points were added to this scatter plot, which of these would be considering an outlier?
   a. (2,29)
   b. (5,20)
   c. (15,4)
   d. (25,5)

435. Which statement is NOT true about this graph?
   a. The data shows a negative linear association.
   b. The data shows that the more hours one studies each week, the more quickly one passes the GED test.
   c. The data shows a negative nonlinear association.
   d. One can quite confidently predict that a student who studies 19 hours a week will pass the GED test in roughly 11 weeks.

Answers and Explanations

382. **The correct answer is choice c.** This graph intercepts the y-axis at -5. It also increases from $(-\infty,-5) \cup (-5,\infty)$.

Choice **a** is incorrect because it decreases on the given interval and has a y-intercept of $+5$ and not -5.

Choice **b** is incorrect because although it is increasing on the given interval, it has a y-intercept of $+5$ and not -5.

Choice **d** is incorrect because although it has a y-intercept of -5, it decreases on the given interval.

383. **The correct answer is choice c.** Functions are decreasing over the intervals where their slope is negative, or going down from left to right. Intervals are defined by the x-values over which a behavior is being exhibited. In this function, the slope is negative from when $x = -1$ until $x = 1$. So this function is decreasing over the intervals $(-1,1)$. Answer choice **d** names the y-values over which the function is decreased, but this is not the correct convention for discussing intervals. Choices **a** and **b** show the intervals over which the function is *positive* and *negative*, respectively (parentheses and brackets should be used in choice **a**).

384. **The correct answer is choice b.** Even degree functions have similar end behavior in both directions, while odd degree functions have opposite behavior in both directions. Since this function is going in two opposite directions as x approaches ∞ and as x approaches $-\infty$, it must be an odd degree function. This rules out choice **a**.

Choice **c** does not make sense since that is a linear equation and this graph is curved.

Choice **d** does not make sense since x is in the bottom of a fraction.

Choice **b** is the correct answer because it shows a cubic function, and cubic functions always have opposite behavior in both directions.

385. **The correct answer is choice c.** The relative minimum of $g(x)$ will be where it hits a valley and the slope has changed directions. Choice **d** shows the lowest point on the graph, but this is not by definition the *relative minimum*.

Choice **b** shows the origin, which is not in a valley.

Choice **a** shows the relative maximum, which is at a peak.

Choice **c** is the relative minimum, and the valley is at $(1,-6)$.

386. **The correct answer is choice d.** A function is positive when its y-value is positive and its graph is above the x-axis. Intervals are expressed in terms of the x-values for which a certain behavior is exhibited. The function $g(x)$ is positive as x goes from -2 to 0 and then again as x goes from 2 to ∞. This is written as $(-2,0)$ and $(2,+\infty)$, so **d** is the correct answer.

Choice **c** is incorrect because it is naming the intervals of the function where it is *increasing* rather than naming where it is *positive*.

Choice **b** is naming just one interval where the function is positive but it is forgetting the interval $(2,+\infty)$.

Choice **a** is listing the height range of the function in the second quadrant, which has nothing to do with the interval of the function being positive.

387. **The correct answer is $84.** Use the bar graph to find how much Cecilia spent on her cell phone in April and how much she spent on her cell phone in May. Then add those two values to find how much she spent in the two months combined.

Each tick mark on the vertical axis of the graph represents $3.

Cecilia spent $54 in April and $30 in May.

$$\$54 + \$30 = \$84$$

Cecilia spent $84 on her cell phone in April and May combined.

388. **The correct answer is choice c.** Looking at the graph, we can consider each answer choice to see which one best approximates the average cost of Cecilia's cell phone bill each month.

Choice **a** doesn't make sense since February is the only bill that is less than $15 and the rest of the months are considerably higher than that.

Choice **b** doesn't make much sense either since only three of the nine months are $20 or less and the remaining six months are nearly $30 or above.

Choice **d** doesn't seem like the best answer since only two out of the nine months are $40 or above; this average seems too high. After eliminating these three choices, it is clear that choice **c** is the best approximation since several of the nine months are reasonably close to $30.

389. **The correct answer is choice c.** Each point on a box plot represents the beginning and/or end of a quartile. The left-most point represents the beginning of the first quartile, the next point represents the end of the first quartile and the beginning of the second quartile, and so on. Each quartile accounts for 25% of the data. Since there are four quartiles in a data set, the middle two quartiles (the second and third quartiles) represent the middle 50% of the data. For this data set, the beginning of the second quartile is $50 and the end of the third quartile is $90. So, the middle 50% of the data is between $50 and $90.

Choice **a** is incorrect because $30 < p < $85 describes the *bottom* 50% and not the middle 50%.

Choice **b** is incorrect because $50 < p < $85 describes the data points that are in the 25th to 50th percentiles, instead of the data that spans from the 25th to 75th percentiles.

Choice **d** is incorrect because $85 < p < $90 describes the data points that are in the 50th to 75th percentiles, instead of the data that spans from the 25th to 75th percentiles.

390. **The correct answer is choice a.** Since $85 represents the median and $90 represents the upper quartile, we can conclude that 25% of the data falls between $85 and $90.

Choice **b** is not correct because 50% of the data is either the range demonstrated by the box ($50 to $90), the range from the minimum to the median ($30 to $85), or the range from the median to the maximum ($85 to $110).

Choice **c** is incorrect because 75% of the books are lower than $90 but higher than $30, but the books between $85 and $90 are not 75% of the data.

391. **The correct answer is 9%.** To find the difference between food expense and the combined total of transportation and clothing expenses, look at the numbers on the graph. Food expense is 22%, transportation is 9%, and clothing is 4%; $22 - (9 + 4) = 9\%$.

392. **The correct answer is 30%.** In a pie chart, all of the percentages must sum to 100%. When adding all the other given percentages together, we get 70%, so this means that 30% is left for housing.

393. **The correct answer is choice a.** When looking at a scatter plot of data points, a correlation exists if there is a relationship between x and y that holds true for the majority of the points. For example, if the y-values get larger as the x-values get larger, there is a positive correlation. In other words, if the points seem to rise as you move from left to right on the graph, there is a positive correlation. Similarly, if the y-values get smaller as the x-values get larger, there is a negative correlation; the values are behaving in an opposite way to each other. According to this scatter plot, there is a distinct relationship. As the x-values get larger, so do the y-values. Therefore, there is a positive correlation.

Choice **b** is incorrect because if there were a negative correlation, the points would be sloping downward to the right, not upward to the right.

Choice **c** is incorrect because if there were no correlation, the points would not be clustered together as they are on this scatter plot.

Choice **d** is incorrect because the values are not necessary when determining if a positive or negative correlation exists in a scatter plot.

394. The scores are added to the dot plot by placing repeated dots over the value on the scale. The circled dots represent the four added students.

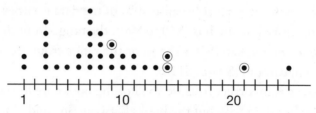

Number of Species Observed

395. The correct answer is choice d. The bar representing distances from 1 up to 10 miles has a height of 3, meaning 3 students reported traveling less than 10 miles to campus.

Choice **a** is incorrect. The total number of students surveyed can be found by adding the frequencies. There were 3 + 6 + 4 + 2 + 1 = 16 students surveyed.

Choice **b** is incorrect. While there was only one student who travels between 37 and 46 miles to campus, there is no way to tell the exact number of miles he travels using this graph.

Choice **c** is incorrect. The 10 and 19 on the horizontal axis represent distance, not frequency.

396. The correct answer is choice a. The slope of the line represents the unit rate. Using the start point (0,50) and the end point (8,250), the slope is $\frac{250 - 50}{8 - 0} = \frac{200}{8} = 25$.

Choice **b** is incorrect. The point 50 on the vertical axis represents the number of packages loaded at the beginning of the day.

Choice **c** is incorrect. There were 125 packages loaded in a little after 3 hours, so it cannot be the hourly rate.

Choice **d** is incorrect. This cannot represent the hourly rate because there were 250 packages loaded after 8 hours.

397. The correct answer is 2002. The largest difference is indicated by the bars representing the enrollment in one year having the largest discrepancy in height. In 2002, the program had an enrollment of approximately 34 male students and 8 female students. This is the largest height discrepancy shown on the graph.

398. **The correct answer is choice a.** There are 5.5 house symbols used in the chart for San Diego, indicating 5.5 × 2,500 = 13,750 subscribing households in that city.

Choice **b** is incorrect. There are 5.5 house symbols, not 6 (which would result in 15,000 subscribing households).

Choice **c** is incorrect. This is the number of subscribing households in Sacramento.

Choice **d** is incorrect. This is the number of subscribing households in San Francisco.

399. **The correct answer is a vertical line at 3.**

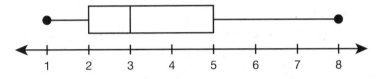

Hours Spent on Computer per Day

A box plot or box-and-whisker plot is a type of graphic that shows five different benchmarks within a set of data: the minimum value, the maximum value, the upper quartile, the lower quartile, and the median. Looking at the given graph, we see that the minimum value of 1 and the maximum value of 8 are illustrated clearly. The median of a data set is the middle piece of data—there is an equal number of data points above and below it. The median is represented by the bar inside the box of the graph. Since 50% of the employees spent more than 3 hours on their computers, 3 is the median and there should be a center bar in the box at 3. The most frequent response from employees of 4 hours is the mode, but the mode is not a statistic that is represented in box-and-whisker plots. The mean time spent on computers of 4.5 hours is also not displayed since the arithmetic average is not shown in box-and-whisker plots. Once the additional bar showing the median is added, you can see that the overall distribution of time employees spent on their computers is skewed right.

400. **The correct answer is choice a.** The revenue is increasing whenever the graph is rising from left to right. This occurs between weeks 2 and 3.

Choice **b** is incorrect. The revenue is increasing whenever the graph is rising from left to right. This does not occur between weeks 3 and 4.

Choice **c** is incorrect. The revenue is increasing whenever the graph is rising from left to right. This does not occur between weeks 4 and 5.

Choice **d** is incorrect. The revenue is increasing whenever the graph is rising from left to right. This does not occur between weeks 6 and 7.

401. **The correct answer is choice a.** The pattern in the scatter plot has a general upward trend from left to right. This indicates a positive relationship. As one variable increases, the other variable also increases.

Choice **b** is incorrect. A negative relationship would be indicated by a pattern that is generally falling from left to right.

Choice **c** is incorrect. This would be true if, for each point, the x- and y-coordinates were the same. But there are many points where this is not the case.

Choice **d** is incorrect. A general sloping pattern indicates a relationship between the two variables.

402. **The correct answer is choice c.** Since product 2 had almost $30,000 in sales and product 4 had over $40,000 in sales, the total must be more than $50,000.

Choice **a** is incorrect. The total sales in July for these two products were about $45,000.

Choice **b** is incorrect. The total sales in July for these two products were slightly less than $50,000.

Choice **d** is incorrect. The total sales in July for these two products were about $35,000.

403. The correct answer is choice b. The mode is the number that occurs most often in a set of data. The bar graph shows that in this town, families having two children are most numerous.

Choice **a** is incorrect because this corresponds to families with one child, who are not as numerous as families having two children, and therefore this is not the correct answer.

Choice **c** is incorrect because this corresponds to families with three children, who are not as numerous as families having two children, and therefore this is not the correct answer.

Choice **d** is incorrect because this corresponds to families with four children, who are not as numerous as families having two children, and therefore this is not the correct answer.

404. The correct answer is choice b. First, add up all the values: $10 + 10 + 5 + 5 + 10 = 40$. Next, divide by 5 (the number of values): $40 \div 5 = 8$ inches.

405. The correct answer is 63. A box-and-whisker plot is a type of graphic that shows five different benchmarks within a set of data: the minimum value, the maximum value, the upper quartile, the lower quartile, and the median. The median of a data set is the middle piece of data—there is an equal number of data points above and below it. The median is represented by the bar inside the box of the graph, which is at 63 in this case.

406. The correct answer is choice b. Half of the survey responses were "dog." Therefore, half of the circle graph will be the dog sector. The entire circle graph has 360°. Half of 360° is 180°. The dog sector should be 180°.

407. The correct answer is choice c. To find the mean of the numbers given, add the four numbers and divide by 4: $28 + 44 + 64 + 56 = 192$; $\frac{192}{4} = 48$.

408. The correct answer is choice c. Find the slope of the graph to find how many miles per hour Rhonda was biking. Identify two points on the graph to $\frac{\Delta y}{\Delta x}$. There are several points on this graph, but one point is approximately (6,42) and another point is approximately (2,14). Therefore, $\frac{\Delta y}{\Delta x} = \frac{42-14}{6-2} = \frac{28}{4} = 7$ mph. Choices **a** and **b** are incorrect; even if the points are rounded slightly differently than described, these estimates are too low. Choice **d** is incorrect because the estimate is too high.

409. **The correct answer is choice c.** The description suggests that the plant and rabbit populations decrease and increase together, which is indicated by the graph.

Choice **a** is incorrect. The graph suggests that the plants and rabbits have an inverse relationship, which is not true.

Choice **b** is incorrect because there is no cyclic behavior present in the graph; it suggests that the plant population dies and the rabbits continue to flourish.

Choice **d** is incorrect because the graph suggests that both the plant and rabbit populations remain constant over time, which is not suggested in the description.

410. **The correct answer is choice d.** Use the two labeled points to determine the slope of the best fit line is $\frac{8-4}{28-12} = \frac{1}{4}$. Using the point (12,4) with this slope, we can write the equation of the best fit line in slope-intercept form as $y - 4 = \frac{1}{4}(x - 12)$, so that $y = \frac{1}{4}x + 1$. Now we must determine the value of x for which $y = 10$: $\frac{1}{4}x + 1 = 10$, so that $x = 36$.

Choice **a** is incorrect because you divided the desired rating by the slope of the best fit line, but this is not how you use the best fit line to determine a rating.

Choices **b** and **c** are incorrect because 30 and 32 are too low, as can be seen on the best fit line.

411. **The correct answer is choice d.** Forty people were asked the question and 11 of them responded that they prefer receiving a text message. This gives $\frac{11}{40} = 27.5\%$. So, you would expect 27.5% of 8,000 = 2,200 alumni to prefer receiving a text message.

Choice **a** is incorrect because this is the number you would expect to prefer e-mail.

Choice **b** is incorrect because this is the number you would expect to prefer a newsletter.

Choice **c** is incorrect because this is the number you would expect to *not* prefer a text message.

412. **The correct answer is choice d.** The diagonals are congruent. Use the Pythagorean theorem to determine the length of $\overline{SQ}$: $\sqrt{a^2 + x^2}$. Double this to get the desired sum.

Choice **a** is incorrect because the lengths of the diagonals of a rectangle are the same, so they should be expressed using the same expression.

Choice **b** is incorrect because the length of each diagonal is given by the Pythagorean theorem, which requires a square root.

Choice **c** is incorrect because this is the perimeter of the rectangle, not the sum of the lengths of the diagonals.

413. **The correct answer is choice b.** Since more than 50% of those polled preferred this song, this is a reasonable conclusion.

Choice **a** is incorrect because while this is true about the sample, it does not have to be true of the entire audience.

Choice **c** is incorrect because while this is true about the sample, it does not have to be true of the entire audience.

Choice **d** is incorrect because the song could be a very well-liked older song.

414. **The correct answer is choice a.** The portion of the findings in this scavenger hunt that is quarters is 12/40, and the portion that is silver dollars is 4/40. Adding these, you would expect that 16/40, or 40%, of the findings from a similar scavenger hunt would be composed of quarters and silver dollars.

Choice **b** is incorrect because this is the portion of the findings that you would expect to be quarters.

Choice **c** is incorrect because this is the portion of the findings that you would expect to be silver dollars.

Choice **d** is incorrect because this is the portion of the findings that you would expect to be pennies.

415. **The correct answer is choice b.** Use the two labeled points (1,9) and (3,7) to get that the slope of the best fit line is –1. The equation of the best fit line is $y - 9 = -(x - 1)$, so that $y = -x + 10$. Thus, $y(4) = 6$. This means that you would expect to make 6 errors if you had 4 hours of sleep.

Choice **a** is incorrect because this is the y-intercept of the best fit line and would represent the number of errors you would expect to make based on 0 hours of sleep.

Choice **c** is incorrect because 1 is the slope, but is not the number of errors you would expect to make based on 4 hours of sleep.

Choice **d** is incorrect because this is the number of errors based on 5 hours of sleep.

416. **The correct answer is choice a.** Determine the average of the 40th and 41st ratings out of 79: $\frac{3+4}{2}$ = 3.5 stars.

Choice **b** is incorrect because this is the average of the 41st and 42nd ratings, but the median is the average of the 40th and 41st ratings.

Choice **c** is incorrect because this is the average of the 39th and 40th ratings, but the median is the average of the 40th and 41st ratings.

Choice **d** is incorrect because this is the mean rating, not the median.

417. **The correct answer is choice d.** The difference in the bill for these two months is $240 – $160 = $80, which is the largest of the differences between any two consecutive months.

Choice **a** is incorrect because the difference in the bills for these two months is $40, and other choices produce a larger difference.

Choice **b** is incorrect because there is no difference between the bills for these two months, and other choices do produce a difference.

Choice **c** is incorrect because the difference in the bills for these two months is $40, and other choices produce a larger difference.

418. **The correct answer is choice c.** The mode of a data set is the most
frequently occurring data value, which is 16.
Choice **a** is incorrect because this is the maximum value, not the
mode.
Choice **b** is incorrect because this is the median, not the mode.
Choice **d** is incorrect because this is the range (maximum value –
minimum value), not the mode.

419. **The correct answer is choice c.** Use the two points labeled on
the line to find the slope of the line: $m = \frac{105 - 90}{25 - 30} = -3$. So the equation of the line is $y - 105 = -3(x - 25)$, which is equivalent to
$y = -3x + 180$. So $y(50) = 30$.
Choice **a** is incorrect because this is the average number of text
messages for a 40-year-old.
Choice **b** is incorrect because this is the average number of text
messages for a 30-year-old.
Choice **d** is incorrect because this is the average number of text
messages for a 15-year-old.

420. **The correct answer is (–1,2).** The relative maximum in a graph
will be any point that is the highest point in that section of the
graph. The points immediately to the left and to the right of the
relative maximum will be lower than it, although the relative maximum is not necessarily the highest value on the graph. The given
function has a relative maximum at (–1,2), since it is at the top of
a peak in the graph and is higher than its surrounding points.

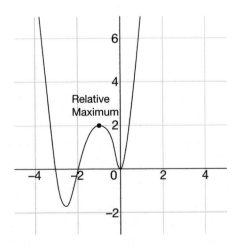

421. **The correct answer is 8.** The first generation displayed is Generation 1. The height of the bar for Generation 1 is 10, which means each household had an average of 10 children. The height of the bar for the current generation, Generation 5, is 2, which means that each household in the Dexter family today has an average of 2 children. On average, the difference between the first generation displayed and the current Dexter generation is 8 children per household.

422. **The correct answer is choice d.** The 5th (current) generation in the Dexter family has 2 children. Since the number of children has gone down from each generation to the next, it would be least likely for the number of children in the 6th generation to *increase* to 3 children (choice **d**). Answer choice **c**, 2 children, is also not likely, since the trend has been for the number of children to *decrease*, but 3 children is more unlikely since increasing the number of children is the opposite of decreasing the number of children in a generation (choices **a** and **b**).

423. **The correct answer is choice b.** In order to calculate what percentage of the books at Everdale Library are biographies, divide the number of biographies by the total number of books. First find the total number of books:

Reference:	750
Mystery:	600
Sci-fi:	250
Children's:	600
Biography:	400
Total:	2,600

Since there are 400 biographies, calculate that $\frac{400}{2,600} \approx 0.154 \approx 15.4\%$.

Choice **a** is incorrect because you calculated the percentage sci-fi books at the Everdale Library.

Choice **c** is incorrect because although there are five different categories of books, and biographies are just one of those categories, computing that 1 in 5 is 20% does not take into consideration that the categories have varying numbers of books.

Choice **d** is incorrect because you calculated the percentage of either mystery or children's books at the Everdale Library.

424. **The correct answer is choice d.** According to the bar graph, the height of the sci-fi bar is at approximately 250 books, while the height of the reference bar shows about 750 books. Therefore, there are about $\frac{1}{3}$ the number of sci-fi books as reference books. Choice **c** is an easy mistake to make because this reverses the relationship: there are three times as many reference books as there are sci-fi books.

425. **The correct answer is choice a.** To represent information in a circle graph, we have to first find the percentage of the whole for each statistic, since the circle represents a whole. First, we need to find the total number of single parents by adding 10.322 million and 1.956 million:

$$\begin{array}{r} 10.322 \text{ million} \\ + \ 1.956 \text{ million} \\ \hline 12.278 \text{ million} \end{array}$$

Now that we know the total number of families, we can find the percentage of single-mother families and single-father families by doing division problems:

$$\frac{\text{\# of single-mother families}}{\text{Total number of families}} = \frac{10.322 \text{ million}}{12.278 \text{ million}} = 0.8406$$
$$= 84.1\% \text{ (when rounded)}$$

$$\frac{\text{\# of single-father families}}{\text{Total number of families}} = \frac{1.956 \text{ million}}{12.278 \text{ million}} = 0.1593 =$$
15.9% (when rounded)

Choice **b** has the statistics for single fathers and single mothers reversed.

Choice **d** is a rounded answer, but with choice **a** available, choice **d** is not as accurate.

Choice **c** is similar to choice **d**, but the statistics are reversed.

426. **The correct answer is choice b.** Looking at the pie chart, Hispanic/Latin origin people make up about one-third of the population. If the pie chart represents a total of 10 million people, then one-third of 10 million would be 3.3 million, so choice **b** is the best answer. Choice **a** is incorrect because 5 million Hispanic/Latin origin people out of a total of 10 million would be 50% of the population and the graph does not show Hispanic/Latin origin people taking up half the circle chart. Similarly, choices **c** and **d** represent one-fifth and one-tenth of the population, but by looking at the graph, we can see that the Hispanic/Latin origin people make up more the one-fifth of the circle graph.

427. **The correct answer is choice b.** Looking at the circle graph, we can see that just under 50% of the population in this country is white, so answer choice **b** is the correct answer.

Answer choices **c** and **d** would require the section representing whites to take up more than half the graph, which it doesn't. Choice **a** is way too low to be correct.

428. Since the range of data went from 0 to 14, make a number line that is evenly spaced from 0 to 14. Each time a data point appears in the set, put an x above that data label:

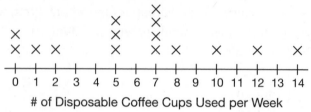

# of Disposable Coffee Cups Used per Week

429. **The correct answers are true and 3.** Looking at the dot plot, it is apparent that four officers use fewer than four disposable cups per week and 11 use more than four cups, so it is true that the majority of the police force use at least four disposable coffee cups per week. In the second sentence, the phrase *fewer than two disposable coffee cups* includes 0 coffee cups and 1 coffee cup, and these two options have a total of three responses indicated.

430. **The correct answers are Alaska and 25%.** The span of the daily high temperature in Minnesota goes from 47 degrees to 76 degrees, whereas the daily high temperature in Alaska goes from 53 degrees to 70 degrees. The smaller range of temperatures in Alaska illustrates that it has a more consistent temperature in the fall. Looking at the box-and-whisker plot for Minnesota, the left-hand side of the box represents the lower quartile, which indicates that 25% of the days have a daily high of no more than 62 degrees. Even though the 15-degree span of the 47-degree minimum to the 62-degree lower quartile accounts for about half of the 29-degree range in temperatures, the lower whisker of a box-and-whisker plot represents the bottom 25% of the data points.

431. **The correct answer is choice c.** Determine how many students are in each of the different age ranges and find the sum:

41- to 45-year-olds: 9 students

46- to 50-year-olds: 12 students

51- to 55-year-olds: 11 students

56- to 60-year-olds: 7 students

Total = 39 students

432. **The correct answer is 31%.** To find the percentage of students ages 46 to 50, you first have to find the total number of students. In question 431, we calculated 39 total students. Reading the graph, we can see that the number of students ages 46 to 50 is 12. To find the percentage, we simply divide 12 by 39 and then move the decimal point two spaces to the right.

$$\frac{12}{39} = 0.307 = 30.7 = 31\% \text{ (when rounded)}$$

So, 31% of Mr. Duvall's students are ages 46 to 50

433. **The correct answer is 13%.** We found the percentage of the age group with the highest number of students in the previous question. Next, we need to find the percentage of students in the age group with the lowest number of students. The lowest number of students in an age group is 7, in the 56 to 60 age range. Find the percentage by dividing 7 by the total number of students, and then move the decimal place.

$$\frac{7}{39} = 0.179 = 17.9 = 18\% \text{ (when rounded)}$$

The percentage difference is then

31%

−18%

13%

So, the percentage difference between the age groups with the highest number of students and the lowest number of students in Mr. Duvall's class is 13%.

434. **The correct answer is choice c.** An outlier is point on a scatter plot that is located away from the cluster and the trend line. Looking at this scatter plot one would imagine that a student studying 15 hours a day would need anywhere from 10 to 16 weeks to prepare to pass the GED test. The point (15,4) indicates a much lower number of weeks needed than what the trend indicates, so it is an outlier.

The points in choices **a** and **b** would blend in with the existing points and would not be considered outliers.

Although the point illustrated in choice **d** would be farther to the right of the rest of the points on the scatter plot, it would be close to a trend line drawn to model this data.

435. **The correct answer is choice c.** The graph *does* show a linear association, so choice **c** is not true.

Choice **a** is an incorrect choice because the data *does* show a negative linear association.

Choice **b** is an incorrect choice because as the number of hours of studying increases, the number of weeks it takes to pass the GED test decreases.

Choice **d** is an incorrect choice because since the graph does have a linear association, one can predict values not plotted that follow the linear progression.

9

Statistics and Probability

The field of statistics revolves around data—how it is collected, organized, and manipulated so that we can interpret information to make predictions. Understanding measures of central tendency, like mean, median, and mode, is important for being able to reduce a body of information into a simpler summary number. Probability is an application of statistics aimed at predicting the likelihood of events happening. Last, counting methods exist to help us calculate the number of different groupings that are possible from a given set of options. In this chapter you will be tested on your fluency with the following concepts:

- Mean, median, and mode of data sets

- Weighted averages

- Simple probability

- Compound probability

- Permutations

- Combinations

436. What is the mode of the following data set? {45, 56, 23, 36, 45, 79, 12, 12, 56, 38, 80}
  a. 79
  b. 45
  c. 12, 45, 56
  d. 80

437. Phoebe works part-time at the movie theater. Her schedule for the next three weeks, shown here, lists the number of hours Phoebe will work each day.

SUNDAY	MONDAY	TUESDAY	WEDNESDAY	THURSDAY	FRIDAY	SATURDAY
0	7	0	4	4	5	0
0	4	6	5	3	2	0
0	5	4	3	6	5	0

What is the median number of hours Phoebe will work in one day over the next three weeks?
  a. 0 hours
  b. 3 hours
  c. 4 hours
  d. 5 hours

438. Pat spends Friday night at the bowling alley. In his first four games, he bowls scores of 123, 165, 127, and 144. If the mean of Pat's five games is 146, what does Pat bowl in his fifth game?
  a. 152
  b. 156
  c. 168
  d. 171

439. Which of the following data sets has a mean of 6, median of 7, and mode of 2?
a. 2, 2, 2, 2, 7, 8, 8, 8, 10
b. 2, 2, 6, 9, 15
c. 2, 2, 2, 7, 7, 11, 11
d. 2, 6, 6, 6, 15

440. Mr. Carlo's class is learning about frogs. He took his class to the pond to observe them in their environment. One of the assignments was for each of the 14 students to measure one frog in order to calculate an average length. The average length was 2.35 inches. Using the following data, calculate the length of the 14th frog.

FROG	LENGTH (IN.)
1	2.3
2	1.9
3	2.0
4	2.4
5	2.5
6	3.0
7	2.7
8	2.6
9	2.5
10	2.4
11	2.3
12	2.1
13	2.4
14	x

441. Holly is competing in four different gymnastics events and wants to get an average score of 9.2. Her vault score was 8.9, her uneven bar score was 8.6, and her balance beam score was 9.5. What must she score on her floor routine to achieve her goal of a 9.2 average for the meet? _____

442. Yolanda is playing a memory card game with her niece. They have a stack of 28 cards made up of 14 pairs of matching animals. Each player gets a turn to flip over two cards in hopes of finding a match. If Yolanda goes first, what is the probability that she will get a matching pair during her first turn?

a. $\frac{1}{27}$

b. $\frac{1}{28}$

c. $\frac{1}{378}$

d. $\frac{1}{756}$

443. How many different ways could the first-, second-, and third-place trophies be awarded to the Little League teams in the end-of-season tournament?

Panthers
Cougars
Sharks
Lions
Tigers
Blue Devils
Mariners

a. 3

b. 35

c. 210

d. 5,040

444. Joan has seven CDs she wants to pack in her suitcase, but only four will fit. How many different combinations of CDs could be packed in Joan's suitcase?

a. 28

b. 35

c. 210

d. 840

445. The graph shows how much homework Michael has done each night. What is the mean number of hours Michael has spent doing homework on the nights shown? _____

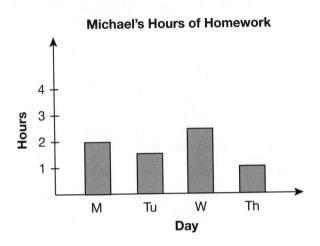

Michael's Hours of Homework

446. Mr. Kissam has the following seven final exam scores in his grade book for his Art History class: 92, 84, 79, 92, 84, 94, and 92. There were eight students in the class, but he forgot to record the final exam grade for the eighth student before passing the exams back. Mr. Kissam sees this student a few days later at the pizza shop and asks what his final exam score was. The student couldn't recall exactly but he told Mr. Kissam that he did better than his friend who got an 84 but not as good as his roommate, who scored a 92. If Mr. Kissam recalls that the median of all eight scores was 90.5, find the missing student's grade. _____

447. A basketball coach has 9 players on her team. How many different 5-player lineups can she create for the starting team?
a. 15,120
b. 45
c. 126
d. 25

448. A teacher would like to pick two students from her class of 30 (16 girls and 14 boys) to be class leaders. If she picks these students one at a time, without replacement, what is the probability that both class leaders are boys? Round your answer to the nearest whole percent.

    a. 14%

    b. 21%

    c. 47%

    d. 91%

449. What is the mode of the data set 9, 4, –1, 12, 4, 8, 7?

    a. –1

    b. 4

    c. 7

    d. 13

450. There are 48 total applicants for a job. Of these applicants, 20 have a college degree, 15 have five years of work experience, and 8 have a college degree and five years of work experience. If an applicant is randomly selected, what is the probability, to the nearest tenth of a percent, that he or she has a college degree or has five years of work experience?

    a. 41.7%

    b. 56.3%

    c. 72.9%

    d. 89.6%

451. Myrna teaches at a university. When assigning final grades, she gives each type of assignment a different level of importance. The collection of Myrna's assignments is worth a total of 100 points.

ASSIGNMENT	NUMBER	PERCENTAGE
Tests	4	70%
Homework	6	10%
Final exam	1	20%

Chad's scores are shown in the next table. What is his final grade for the course? _____

ASSIGNMENT	SCORES
Tests	78, 85, 88, 90
Homework	87, 90, 83, 93, 91, 90
Final exam	82

452. Ethan writes down the sale of every ice cream cone by flavor during his shift at Ice Cream Heaven: vanilla, chocolate, strawberry, strawberry, coffee, chocolate, pistachio, strawberry, mint, vanilla, mint, chocolate, butter pecan, chocolate, coffee, pistachio, chocolate.
Which ice cream flavor is the mode during Ethan's shift?
a. chocolate
b. butter pecan
c. strawberry
d. vanilla

453. Louis has a jar full of jelly beans: six lemon, four orange, three licorice, five lime, and two grape. If he selects one jelly bean at random, what is the probability that it will NOT be lemon?
a. $\frac{1}{20}$
b. $\frac{3}{10}$
c. $\frac{7}{10}$
d. $\frac{7}{20}$

454. Annabelle is holding six cards: three spades, one heart, and two clubs. Finn will select two cards in a row out of Annabelle's hand. What is the probability that he will choose a heart followed by a spade?

a. $\frac{1}{10}$

b. $\frac{1}{36}$

c. $\frac{1}{3}$

d. $\frac{1}{6}$

455. A pair of number cubes is rolled. Let A be the event that both number cubes show the same number. Let B be the event that the sum of the numbers showing on both number cubes is odd. What is the probability that both events A and B occur on the same roll of the number cubes?

a. 0

b. $\frac{1}{6}$

c. $\frac{1}{2}$

d. $\frac{2}{3}$

456. Jessalyn looks at the children's menu in Harper's Restaurant. She has a choice of three appetizers, six main courses, and four desserts. If Jessalyn orders a meal consisting of an appetizer, a main course, and a dessert, how many different meal combinations are available to her?

a. 13

b. 24

c. 72

d. 720

457. Ash is a real estate agent trying to sell a home in a neighborhood she's not familiar with. She wants to be able to tell the family she's representing what the median home price is for the local sales that have occurred over the past month. The past month, housing sold for the following prices: $280,000, $200,000, $424,000, $390,000, $280,000, and $320,000. Find the median house price. _____

For questions 458 to 461, use the following bar graph, which illustrates the number of pencils students had in their bags when arriving to an art class.

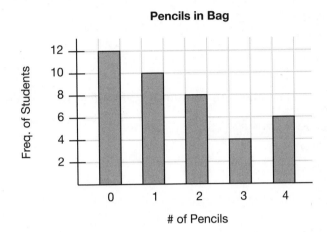

Pencils in Bag

458. What was the median number of pencils that students had in their bags when they arrived to art class?
  a. 0
  b. 1
  c. 2
  d. 4

459. What number of pencils represents the mode?
  a. 0
  b. 1
  c. 2
  d. 4

460. If a student is chosen at random from the class, what is the probability that he or she will have exactly 3 pencils? _____

461. If a student is chosen at random from the class, what is the probability that he or she will have at least 1 pencil? _____

462. What is the approximate surface area of the kickball that has a radius of 4.5 inches?

a. 60 in.²

b. 64 in.²

c. 254 in.²

d. 300 in.²

463. Refer to the table with the number of candies shown here:

Green	6
Yellow	5
Brown	10
Red	8
Blue	7
Orange	6

What is the percentage probability of grabbing a blue candy the first time and then another blue candy the next time, without replacing the first blue candy? _____

464. Mitch has six songs he would like to add to his portable music player, but only three will fit. How many different three-song groupings does Mitch have from which to choose? (*Hint:* Order does NOT matter.)

a. 18

b. 20

c. 120

d. 720

465. Ten people were asked to state their favorite number, and the results are shown in the table.

PERSON	A	B	C	D	E	F	G	H	I	J
Favorite number	7	3	7	1	100	360	2	17	100	7

What is the mode in the list of favorite numbers?

a. 360

b. 100

c. 7

d. 1

466. Kyra has a bag of jelly beans. The bag contains 7 red, 3 orange, 3 black, and 8 purple jelly beans. If her friend randomly picks one jelly bean from the bag, what is the probability that it is red?

   a. $\frac{7}{20}$

   b. $\frac{4}{21}$

   c. $\frac{1}{3}$

   d. $\frac{1}{2}$

467. The chart shows the colors of replacement parts for a popular model of cell phone. The total number of parts shipped is 1,650.

BOXED SET OF REPLACEMENT PARTS	
PART COLOR	**NUMBER OF PIECES**
Green	430
Red	425
Blue	x
Yellow	345
Total	1,650

If a person randomly grabbed a part out of the box, what is the probability that the part will be blue?

   a. $\frac{1}{4}$

   b. $\frac{1}{9}$

   c. $\frac{1}{12}$

   d. $\frac{3}{11}$

Use the following information to answer questions 468–470:

Every student in the senior class at Alexa Mae prep school gets assigned a unique three-digit ID number between 000 and 999. All of the ID numbers have three digits, so ID number 7 is written 007 and ID number 16 is written 016. Once a number is assigned to a student, it cannot be assigned again.

468. Delilah is obsessed with π, so she is hoping to get the number 314. What is the probability that she'll get 314 as her ID number?

469. Skye loves triple repeating digits and doesn't care which number it is, as long as her ID number is three of the same number. What is the probability that Skye will get a repeating digit ID number?

470. Lulu has her heart set on any number with consecutive digits counting up or down, such as 123 and 876. What is the probability she'll get a number she wants?

471. A movie theater sells a special combo during the afternoon matinee. Moviegoers can choose one main item, one candy, and one drink. The options are shown in the table. How many different combos are possible?

MAIN ITEM	CANDY	DRINK
Hot dog	Tasty tarts	Cola
Nachos	Chocolate bites	Diet cola
Large popcorn	Chewy fruits	Iced tea
		Orange soda
		Bottled water

a. 11
b. 30
c. 45
d. 50

472. Ms. Jay has 23 students in her kindergarten class. Every Friday, she recognizes five students who behaved really well all week. Their reward is to be the first in line for lunch the entire following week. How many different combinations of five students are there who could earn this honor?

a. 5
b. 120
c. 33,649
d. 4,037,880

473. The following table gives the ages of the attendees of an awards banquet:

AGE	FREQUENCY
10 to 15	8
16 to 21	31
22 to 30	11
31 to 40	48
41 to 50	32
51 and up	20

What is the probability that the age of a randomly selected guest is greater than 40?

a. $\frac{13}{25}$

b. $\frac{2}{15}$

c. $\frac{49}{75}$

d. $\frac{26}{75}$

474. Several singles tennis games are played in a given day at a weekend tournament. The following are the number of ace serves scored by various players in their matches that day:

NUMBER OF ACE SERVES	FREQUENCY
0	5
2	4
4	2
6	1
8	2
10	4
12	6

What is the median number of ace serves scored?

a. 7

b. 6

c. 8

d. 12

475. Mr. Gallespie is a high-school science teacher. The weight he gives each assignment is listed in the following table.

ASSIGNMENT	NUMBER	PERCENTAGE
Tests	5	60%
Homework	5	15%
Experiments	4	25%

Looking at Katie's scores, calculate her final grade for the semester to the nearest hundredth of a percentage point. _____

ASSIGNMENT	SCORES (OUT OF 100)
Tests	93, 97, 88, 91, 95
Homework	90, 99, 100, 95, 96
Experiments	89, 90, 85, 92

476. A starting basketball player scores the following numbers of points in a series of playoff games:

48, 38, 38, 34, 44, 44

How many points must he score in a seventh game to ensure that the mean number of points scored for all seven games is 40?
a. 41
b. 34
c. 44
d. 38

477. Suppose the probability of a spinner with red and green spaces landing on green is $\frac{2}{5}$. If you spin the spinner once, and then a second time, what is the probability of it landing on green on the second spin?

a. $\frac{3}{5}$

b. $\frac{1}{2}$

c. $\frac{4}{25}$

d. $\frac{2}{5}$

478. Downtown Wheels has different varieties of bikes available and on sale. There are four colors (red, green, black, and silver), two types of seats (large and standard), and three types of structures (beach cruiser, mountain bikes, and 10-speed). If a selection is drawn at random, what is the probability that it is a mountain bike?

a. $\frac{1}{24}$

b. $\frac{1}{12}$

c. $\frac{1}{6}$

d. $\frac{1}{3}$

479. Jackie plays a ring-finding game in her swimming pool. She has 30 seconds to retrieve as many rings as she can from the bottom of the pool. She plays the game nine times, and her scores are shown in the table. What is Jackie's median score?

GAME NUMBER	RING SCORE
1	4
2	3
3	5
4	8
5	6
6	3
7	9
8	3
9	4

a. 3

b. 4

c. 5

d. 6

480. The final scores on a 60-second run of a video game played by seven friends are 168, 121, 210, 96, 215, 118, 320. It was determined later that the score recorded as 320 should have been 350. Which of the following statements is true?

  a. Both the mean and median of the modified data set will increase.

  b. The mean and median of the modified data set will remain unchanged.

  c. The median of the modified data set will decrease, but the mean will increase.

  d. The median of the modified data set remains unchanged, but the mean increases.

481. The probability of pulling a fork out of a drawer is $\frac{4}{13}$. A utensil is removed and placed on the counter. Then another utensil is removed from the same drawer. What is the probability that both utensils drawn from the drawer were forks? _____

482. Wyeth would like to pick three students from his class of 30 students to be class leaders. His class has 16 girls and 14 boys. If Wyeth picks these students one at a time, without replacement, what is the probability that all three class leaders are boys? Express your answer as a percentage to the nearest whole number. _____

483. The following scatter plot shows the average number of steps walked daily by participants in a "Walk for Life" program at a local gym versus the percent success rate for losing weight experienced by them:

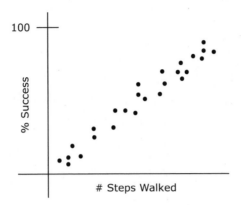

Which of the following best depicts the best fit line for this data?

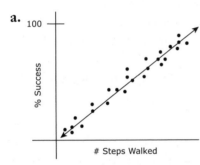

a.

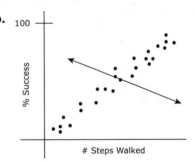

b.

c.

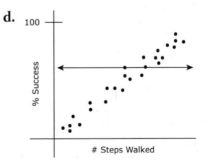

d.

484. An amateur bowler must bowl five games for a ninepin tournament. Her first four scores are 191, 218, 210, and 171. Within what range must her fifth game fall to ensure that her five-game average is between 200 and 215?

　　a. 200 to 215
　　b. 10 to 70
　　c. 210 to 285
　　d. 171 to 218

485. The following lettered tiles (with the number of each indicated beneath each letter) are placed in a bag:

E	P	U	R	B	C
4	2	1	3	2	2

If a single tile is selected at random from the bag, which of the following statements is false?

　　a. It is equally likely to choose a vowel or a consonant.
　　b. There is a 50% chance of selecting a tile that has an R, B, or C on it.
　　c. It is equally likely to choose a tile with a P or B on it.
　　d. It is twice as likely to choose an E as it is to choose a C.

486. A spinner is divided into ten equal sections, numbered 1 through 10. If the spinner is spun once, what is the probability that the spinner will land on a number less than 5?

　　a. $\frac{1}{5}$
　　b. $\frac{2}{5}$
　　c. $\frac{1}{2}$
　　d. $\frac{1}{10}$

487. A piggy bank contains three quarters, five pennies, two nickels, and six dimes. Evander picks a coin at random from the bank and pulls out a quarter. This quarter is NOT replaced. If Evander selects another coin, what is the probability that it will be a quarter?

    a. $\frac{2}{15}$

    b. $\frac{3}{15}$

    c. $\frac{2}{16}$

    d. $\frac{3}{16}$

488. Compute the median of the following list of numbers:
       30, –3, 7, 8, 11, 16, and 22

    a. –3

    b. 11

    c. 13

    d. 16

489. There are 13 different appetizers to choose from at a restaurant. Mahershala and his friends want to order three appetizers for the table to share. How many different combinations of three appetizers could they order? _____

490. If Nadezda flips a coin and rolls a die, what is the probability that she will get tails when she flips the coin and a 3 when she rolls the die? _____

491. The following table illustrates the number of miles each student travels one way to get to Ms. Bradley's viola class. Find the average number of miles that Ms. Bradley's students travel one way:

NUMBER OF MILES TRAVELED ONE WAY	NUMBER OF STUDENTS
1	4
2	3
3	5
4	6
5	3

Round your answer to the nearest tenth of a mile.
a. 3.1 miles
b. 3.7 miles
c. 4.0 miles
d. 4.2 miles

492. Stephanie owns a bakery and is purchasing supplies. If she buys 20 cans of peaches for $6.50 each, and 28 cans of cherries for $9 each, what is the average price per can of fruit? _____

493. Mr. Syed's class is divided into five groups. There are six students in group A, eight students in group B, four students in group C, five students in group D, and four students in group E. If Mr. Syed randomly selects a student to water the plants, what are the odds that he chooses a student from group A?
a. 7 to 2
b. 2 to 9
c. 7 to 9
d. 2 to 7

494. Maya loves to prepare tasty food and host dinners at her home. She has four favorite salads, five favorite entrées, and three favorite desserts that she likes to make. Using these recipes, how many unique combinations can Maya offer her guests of one salad, one entrée, and one dessert? _____

495. A deck contains ten yellow cards, ten green cards, five blue cards, three purple cards, and two red cards. If Isabel selects one card at random, what is the probability that the card is NOT blue?

  a. $\frac{1}{6}$

  b. $\frac{4}{5}$

  c. $\frac{5}{6}$

  d. $\frac{29}{30}$

496. Little Troy has a toy filled with different shapes: squares, circles, and triangles. What is the probability—in lowest terms—that Troy will reach in and grab a square and then a triangle, without replacing the square?

SHAPE	NUMBER
Triangles	9
Circles	7
Squares	6

  a. $\frac{6}{22}$

  b. $\frac{9}{21}$

  c. $\frac{9}{77}$

  d. $\frac{54}{462}$

497. Jacob asked all members of the yearbook staff, who are all juniors, to indicate their favorite subject this year. Here are the results:

SUBJECT	NUMBER FOR WHOM THIS IS THE FAVORITE
English	8
History	3
Math	2
Art	4

Based on these results, Jacob concluded that the favorite subject among all his classmates in the junior class is English. Which of the following offers the best explanation as to why his conclusion may be invalid?

a. Jacob should have included more subjects in the list.

b. The sample is not representative of the entire junior class.

c. Jacob should have also asked juniors from other schools this question.

d. Jacob should have surveyed only students *not* on the yearbook staff.

498. A dog's litter contains nine puppies. There are three black males, three spotted males, two brown females, and one spotted female. What is the probability that a customer randomly selects a puppy from this litter that is either male or spotted?

a. $\frac{1}{9}$

b. $\frac{1}{3}$

c. $\frac{4}{9}$

d. $\frac{7}{9}$

499. Julia is wrapping presents to give to her friends. She has 6 different kinds of boxes, 9 different types of wrapping paper, and 11 different colors of ribbon. How many different combinations of packaging and decorations can Julia use?

a. 26

b. 75

c. 594

d. 693

500. What is the probability that this spinner will land on the number 1?

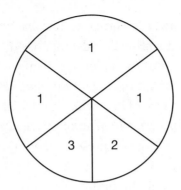

a. $\frac{2}{5}$

b. $\frac{2}{3}$

c. $\frac{1}{3}$

d. $\frac{3}{4}$

501. Suppose you shuffle two standard 52-card decks of playing cards together and select one card at random. What is the probability of randomly selecting either a heart or an ace?

a. $\frac{35}{52}$

b. $\frac{4}{13}$

c. $\frac{17}{52}$

d. $\frac{1}{4}$

Answers and Explanations

436. **The correct answer is choice c.** The *mode* is the number that appears most frequently. This data set has three modes because 12, 45, and 56 each occur twice, more than any of the other numbers. Choice **b** is the correct median of this data set, but not the mode. (It is the central piece of data when it is listed in increasing order.) Choice **a** is the incorrect median of this data set. (It's in the middle of the data set as the data is presented, but the data was not put into increasing order.) Choice **d** is the maximum of the data set but not the mode.

437. **The correct answer is choice c.** The median of a data set is the piece of data that occurs right in the middle after the data is put in order. To find the median number of hours Phoebe will work in one day over the next three weeks, put the number of hours she works each day in order and choose the number in the middle:

0, 0, 0, 0, 0, 0, 0, 2, 3, 3, **4**, 4, 4, 4, 5, 5, 5, 5, 6, 6, 7

There are 21 days on the schedule, so the middle number is the eleventh number shown, 4. The median number of hours Phoebe will work is 4.

Choice **a** is incorrect because you confused the mode with the median. The mode is the most common data entry but the median is the middle data entry when the data is listed from least to greatest.

Choice **b** is incorrect because you confused the mean with the median. The mode is the most common data entry but the median is the middle data entry when the data is listed from least to greatest, but you calculated the mean which is the arithmetic average.

Choice **d** is incorrect because although it is the middle number in the data set as it is listed, the data must first be arranged from least to greatest before determining the median and you forgot this critical step.

438. **The correct answer is choice d.** The mean, or average score, of Pat's five games is 146. That means that Pat scores a total of 5 × 146, or 730, over five games. Let x represent Pat's score in the fifth game. His total score over five games is equal to:

$$123 + 165 + 127 + 144 + x = 559 + x$$

Now, set that total equal to Pat's total score, 730:

$$559 + x = 730$$

Subtract 559 from both sides:

$$x = 171$$

Pat scores 171 in his fifth game.

To check your answer, add Pat's score in each of the five games and divide the sum by 5:

$$123 + 165 + 127 + 144 + 171 = 730$$

The dividend should be equal to Pat's mean, 146:

$$730 \div 5 = 146$$

Choice **a** is incorrect because you found the average of the first four games to be 140 and then calculated the average of 140 and 152 would be 146; however, this gives the last game equal weight as the first four games combined. You incorrectly calculated a weighted average.

Choices **b** and **c** are incorrect because although they will bring Pat's average up considerably, they will not raise it to 171.

439. **The correct answer is choice c.** The mode is the data value that occurs most frequently, which is 2. The median is the middle value when the data is arranged in increasing order, which is 7. The mean is $\frac{2 + 2 + 2 + 7 + 7 + 11 + 11}{7} = 6$.

Choice **a** is incorrect because while this data set has mode 2 and median 7, the mean is not 6 (it is approximately 5.4).

Choice **b** is incorrect because the median is 6 and mean is 6.8.

Choice **d** is incorrect because the mode is 6, median is 6, and mean is 7.

440. **The correct answer is 1.8 inches.** Assign a variable to the missing length, x. Write an equation of the average, and then solve for x:

$$\text{Average/mean} = \frac{\text{Sum of data values}}{\text{\# of data values}}$$

$$2.35 = \frac{2.3 + 1.9 + 2.0 + 2.4 + 2.5 + 3.0 + 2.7 + 2.6 + 2.5 + 2.4 + 2.3 + 2.1 + 2.4 + x}{14}$$

$$2.35 = \frac{31.1 + x}{14}$$

$$2.35(14) = \frac{14(31.1 + x)}{14}$$

$$32.9 = 31.1 + x$$

$$\underline{-31.1 \quad -31.1}$$

$$1.8 = \qquad x$$

The length of the 14th frog is 1.8 inches.

441. **The correct answer is 9.8.** Since Holly is competing in four events, the average will be calculated by adding the score from all four events and then dividing that sum by 4:

$$\text{Average} = \frac{\text{vault} + \text{uneven} + \text{balance} + \text{floor}}{4}$$

Fill in the given information, and use the variable f to represent her floor event score. Work backward to find out what her floor event score must be to get an average of 9.2:

$$9.2 = \frac{8.9 + 8.6 + 9.5 + f}{4}$$

$$9.2 = \frac{27 + f}{4}$$

Here we multiply both sides by 4 to get $27 + f$ alone:

$$9.2(\times 4) = \frac{27 + f}{4}(\times 4)$$

$$36.8 = 27 + f$$

$$\underline{-27 \quad -27}$$

$$9.8 = \qquad f$$

So, in order to achieve her goal, Holly must get a 9.8 on her floor routine today!

442. **The correct answer is choice a.** We are looking for the compound probability of P(animal then matching animal). In order to calculate the compound probability of two related events, we multiply the simple probability of each event happening. Since Yolanda does not care what animal is on the first card she flips over, we use a probability of 1 for the first card. Basically, the P(animal) for the first card she flips over is $\frac{28}{28} = 1$, since it doesn't matter *what* animal is on the first card. Now there are 27 cards remaining and Yolanda is hoping to flip over the *one* card that has the matching animal on it. Therefore, the P(then matching animal) is $\frac{1}{27}$. So multiply the two probabilities together to get the compound probability: P(animal then matching animal) $= 1 \times \frac{1}{27} = \frac{1}{27}$. Therefore, the probability that Yolanda gets any matching pair in the first two flips of cards is $\frac{1}{27}$.

Choice **b** is incorrect because after she flips over the first card, there are only 27 cards left that might have the matching animal, so you forgot to reduce the total number of events from 28 to 27.

Choice **c** is incorrect because you used the probability of the first card as being $\frac{2}{28}$ and the second card as being $\frac{1}{27}$. Using $\frac{2}{28}$ would have been correct if Yolanda was looking for the card of a specific animal, like a panda, but since she was looking for a match of *any* two animals, the probability of the first card being an animal was $\frac{28}{28}$, which is 1.

Choice **d** is incorrect because you used the probability of the first card as being $\frac{1}{28}$ and the second card as being $\frac{1}{27}$. The probability of the first card being $\frac{1}{28}$ was incorrect since picking any animal for the first card is $\frac{28}{28}$ which is 1. Then, the probability of the second card matching the first card is $\frac{1}{27}$.

443. **The correct answer is choice c.** This is a problem involving permutation, which requires the following formula: $\frac{n!}{(n-r)!}$, where n = the total number of options and r = the number of options chosen. Since there are seven teams that could win a trophy, $n = 7$. Only three teams will get trophies, so $r = 3$. When these values are substituted into the equation, we get $\frac{n!}{(n-r)!} = \frac{7!}{(7-3)!}$ $= \frac{7 \cdot 6 \cdot 5 \cdot 4 \cdot 3 \cdot 2 \cdot 1}{4 \cdot 3 \cdot 2 \cdot 1} = 7 \cdot 6 \cdot 5 = 210$.

Choice **a** is incorrect because 3 is the number of trophies being given out, but not the number of different ways that the seven teams could be awarded these trophies.

Choice **b** is incorrect because you used combinations instead of permutations to solve this, but since the order is very important (coming in first place is different from coming in third place), order matters and permutations must be used.

Choice **d** is incorrect because you multiplied $7 \times 6 \times 5 \times 4 \times 3 \times 2 \times 1$, but that calculated the number of ways that all seven ranked places in the tournament could be arranged; the question asked for the number of arrangements of just the first *three* places.

444. **The correct answer is choice b.** Joan can fit four of her seven CDs in her suitcase. It is important to understand that the order in which she chooses the CDs does not matter. The group of CDs A, B, C, and D is the same as the group A, B, D, and C, or D, C, A, and B.

This is a combination problem, so you must use the combination formula to find the answer:

$C(n,k) = \frac{n!}{k!(n-k)!}$, where n is the number of options and k is the number of choices made.

Joan has seven CDs and chooses four. Divide by 4! to ensure that you do not count the same group of four CDs more than once:

$$\frac{7 \times 6 \times 5 \times 4 \times 3 \times 2 \times 1}{(4 \times 3 \times 2 \times 1)(3 \times 2 \times 1)} = \frac{7 \times 6 \times 5 \times 4 \times 3 \times 2 \times 1}{(4 \times 3 \times 2 \times 1)(3 \times 2 \times 1)}$$
$$= \frac{7 \times 6 \times 5}{3 \times 2 \times 1} = \frac{210}{6} = 35$$

Choice **a** is incorrect because you performed 7×4, but this is not how combinations are calculated.

Choice **c** is incorrect because you used permutations instead of combinations to solve this problem and you also solved for taking just three CDs, instead of four CDs.

Choice **d** is incorrect because you used permutations instead of combinations to solve this problem. Since the order in which she packs her four CDs is unimportant, combinations must be used.

445. **The correct answer is 1.75 hours.** The mean is the average. To find the average of four numbers, find the sum of the numbers and divide by 4. Michael spent 2 hours doing homework on Monday, 1.5 hours on Tuesday, 2.5 hours on Wednesday, and 1 hour on Thursday. The sum of this time is 7 hours ($2 + 1.5 + 2.5 + 1 = 7$). Next, divide 7 by 4 ($7 \div 4 = 1.75$). The mean is 1.75 hours.

446. **The correct answer is 89.** The median of a data set is always the very middle number when they are arranged in increasing order. With an odd number of data points, finding the median is easy. However, when you have an even number of data points, there are two middle numbers. The median is then the average of those two middle numbers.

    To find the missing score, we first need to order the data points from least to greatest:

    79, 84, 84, 92, 92, 92, 94

    The problem states that the missing point in the data set is between 84 and 92. So, let's let x represent the missing score and add it to our chronological list:

    79, 84, 84, x, 92, 92, 92, 94

    The middle two numbers are x and 92.

    The problem also states that the average of these two numbers is 90.5. Let's set up an equation and then solve for x:

    $$\frac{x + 92}{2} = 90.5$$
    $$2\left(\frac{x+92}{2}\right) = (90.5) \times 2$$
    $$x + 92 = 181$$
    $$\underline{-92 \qquad -92}$$
    $$x = 89$$

    The missing grade is 89.

447. **The correct answer is choice c.** The order in which the starting players are selected is not significant, so use combination to solve this problem. (Using the combination formula will keep from counting players A, B, C, D, and E as a different team lineup than A, B, C, E, and D.)

$$C(n,k) = \frac{n!}{k!(n-k)!}$$
$$C(9,5) = \frac{9!}{5!(9-5)!}$$
$$C(9,5) = \frac{9!}{5! \times 4!} = \frac{9 \times 8 \times 7 \times 6 \times 5!}{5! \times (4 \times 3 \times 2 \times 1)}$$
$$C(9,5) = 126$$

Choice **a** is incorrect because that is the answer from using the permutations formula.

Choice **b** is just the product of the number of players on the team and the number of players in the starting lineup, which has nothing to do with the number of starting teams.

Choice **d** is incorrect because you likely made an error when simplifying the final part of the calculation when working with factorials.

448. **The correct answer is choice b.** Using the multiplication rule for probability, the probability is $\frac{14}{30} \times \frac{13}{29} \approx 0.21$ or 21%.

Choice **a** is incorrect. Since the students are being selected from the entire class, the denominator should be 30 and not 14.

Choice **c** is incorrect. This represents the probability that one girl is randomly selected. The question is asking for a compound probability of both selected leaders being boys.

Choice **d** is incorrect. The probability of an "and" event should use the multiplication rule, not the addition rule.

449. **The correct answer is choice b.** The mode is the most commonly observed value. In this case, 4 occurs the most number of times.

Choice **a** is incorrect. This is the minimum value of the data set.

Choice **c** is incorrect. This is the median of the data set.

Choice **d** is incorrect. This is the range of the data set

450. **The correct answer is choice b.** Given the final question is about an "or" probability, the correct formula to use is $P(A \text{ or } B) = P(A) + P(B) - P(A \text{ and } B)$, where $P(A)$ stands for the probability of the event A occurring. Applying this here:

$P(\text{degree or five years}) = P(\text{degree}) + P(\text{five years}) - P(\text{degree and five years}) = \frac{20}{48} + \frac{15}{48} - \frac{8}{48} = \frac{27}{48} = 0.5625$

Finally, 0.5625 is equivalent to 56.3%.

Choice **a** is incorrect because you only calculated the probability of the candidate having a college degree, but you did not take into consideration the probability of the candidate having work experience.

Choice **c** is incorrect because you calculated the probability of the candidate having a college degree plus the probability of the candidate having work experience, but you forgot to subtract the probability of the candidate having *both* a degree and work experience.

Choice **d** is incorrect because you calculated the probability of the candidate having a college degree, plus the probability of the candidate having work experience, plus the probability of the candidate having both a degree and work experience, but you needed to *subtract* the probability of the candidate having both experiences, and not add it.

451. **The correct answer is 85%.** To find Chad's final grade, each assignment needs to be weighted differently. First, find the average of each type of assignment and then multiply by the percentage Myrna assigns to it. Then, add the percentages together to find Chad's final grade:

Final grade $= 0.70(\frac{78 + 85 + 88 + 90}{4})$
$+ 0.10(\frac{87 + 90 + 83 + 93 + 91 + 90}{6}) + 0.20(82)$
$= 0.70(85.25) + 0.10(89) + 0.20(82)$
$= 59.675 + 8.9 + 16.4$
$= 84.975$

Chad's final grade is $\approx 85\%$.

452. **The correct answer is choice a.** The mode of a data set is the piece of data that occurs the most often. To find the ice cream flavor that is the mode during Ethan's shift, find the ice cream flavor that was sold more than any other flavor. Add up the number of ice cream cone sales by flavor:

butter pecan: 1
chocolate: 5
coffee: 2
mint: 2
pistachio: 2
strawberry: 3
vanilla: 2

The ice cream flavor that was sold the most during Ethan's shift was chocolate, so it is the mode.

453. **The correct answer is choice c.** First, find the total number of jelly beans in the jar: 6 lemon + 4 orange + 3 licorice + 5 lime + 2 grape = 20 jelly beans.

Next, find the number of jelly beans that are *not* lemon: 4 orange + 3 licorice + 5 lime + 2 grape = 14 jelly beans that are *not* lemon. Louis can select 14 jelly beans that are *not* lemon out of the 20 jelly beans that are in the jar. The probability of Louis selecting a jelly bean that is *not* lemon is $\frac{14}{20}$ or $\frac{7}{10}$.

Choice **a** is incorrect because although Louis is selecting one jelly bean from the jar of 20, the probability of him selecting a jelly bean that is *not* lemon is not $\frac{1}{20}$. There is more than one outcome that will make true the event that Louis selects a jelly bean that is *not* lemon.

Choice **b** is incorrect because this is the probability that Louis *will* choose a lemon jelly bean. Six out of 20, or 3 out of 10, jelly beans in the jar are lemon.

Choice **d** is incorrect because when reducing a fraction, you must divide both the numerator and the denominator by the same number. You might have incorrectly reduced the fraction $\frac{14}{20}$.

454. **The correct answer is choice a.** The probability that he will first select a heart is 1 out of 6. Next, his chances for picking a spade will be 3 out of 5. To find the probability of choosing a heart followed by a spade, just multiply these two probabilities: $\frac{1}{6} \times \frac{3}{5} = \frac{3}{30} = \frac{1}{10}$.

Choice **b** is incorrect because it is the probability Finn will draw a heart, with replacement, 2 times from the bunch.

Choice **c** is incorrect because you might have incorrectly simplified $\frac{3}{30}$ to be $\frac{1}{3}$.

Choice **d** is incorrect because it is just the probability Finn will draw a heart; you must complete the question asked.

455. **The correct answer is choice a.** The two events are mutually exclusive. If the pair of number cubes shows the same number, their sum is even. That is, $1 + 1 = 2; 2 + 2 = 4; 3 + 3 = 6; 4 + 4 = 8;$ $5 + 5 = 10;$ and $6 + 6 = 12$. It is therefore impossible for both events to occur simultaneously, so the probability is 0.

Choices **b**, **c**, and **d** cannot be correct unless it were possible to have the sum of the like faces of two number cubes be an odd number. But the sum of two even numbers is always even, and the sum of two odd numbers is also even. Therefore, no value other than 0 is possible for this experiment.

456. **The correct answer is choice c.** Jessalyn can order one of three appetizers. For each appetizer, she can order one of six main courses, and for each of those, she can order one of four desserts. To find the number of different meal combinations, multiply the number of appetizers by the number of main courses, and then multiply that product by the number of desserts: $3 \times 6 \times 4 = 72$.

Choice **a** is incorrect because 13 is the sum of the number of appetizers, main courses, and desserts. You must multiply these numbers, not add them, to find the number of different meal combinations.

Choice **b** is incorrect because 24 is the product of 6, the number of main courses, and 4, the number of desserts. Twenty-four is the number of main course/dessert combinations Jessalyn can order. To find the number of appetizer/main course/dessert combinations, multiply 24 by the number of appetizers.

Choice **d** is incorrect because 720 is much too high. You would get this answer if you multiplied incorrectly, or found 6! instead of multiplying the numbers of appetizers, main courses, and desserts.

457. **The correct answer is $300,000.** In order to find the median of a set of data, they must first be put in increasing order and the *middle* piece of data will be the median.

$200,000, $280,000, $280,000, $320,000, $390,000, $424,000

In this case, there are two house prices that make up the *middle* of the data, so we must take the average of those: $\frac{280,000 + 320,000}{2}$ = 300,000. So the median house price in this neighborhood is $300,000

458. **The correct answer is choice b.** To find the median number of pencils, we first need to find out the total number of students who were in class that day. Add the frequency for each bar: 12 + 10 + 8 + 4 + 6 = 40 students. The median will be the piece of data that is right in the middle when the individual entries are put in increasing order. Since this is an even number of data entries, the middle piece will be the average of the 20th and 21st piece of data. Use the bars to count the data points: Since there are 12 students who brought 0 pencils and 10 students who brought 1 pencil, we know that the 20th and 21st pieces of data are both going to be in the column of students who brought 1 pencil, so the median is 1 (choice **b**).

Choice **a** is incorrect because you chose the mode, which is the most frequent piece of data, but the median is the middle piece of data, once the data has been organized from least to greatest.

Choice **c** is incorrect because although 2 is in the middle of the bar graph's horizontal axes, there are 40 different pieces of data displayed in the bar graph, and the median will be the average of the 20th and 21st pieces of data, which are in the column for 1 pencil. (There are 12 pieces of data in the first column for 0 pencils, and then there are 10 pieces of data in the second column for 1 pencil, so the 20th and 21st pieces of data are there.)

Choice **d** is incorrect because you chose the maximum number of pencils students had, but not the median.

459. **The correct answer is choice a.** The mode is the most frequent piece of data, and, since the bar for 0 pencils is the tallest, we know that 0 is the mode.

Choice **b** is incorrect because you chose the median number of pencils, but not the mode.

Choice **c** is incorrect because only 8 students had 2 pencils, but the mode is the most frequently reported piece of data, which was 0 pencils, reported by 12 students.

Choice **d** is incorrect because you chose the maximum number of pencils students had, but not the mode.

460. **The correct answer is $\frac{1}{10}$.** Looking at the graph, there are 4 students who had 3 pencils in their bags. Since there were 40 students in total (see explanation for question 458 for student total), the probability that a student chosen at random will have exactly 3 pencils is $\frac{4}{40} = \frac{1}{10}$.

461. **The correct answer is $\frac{7}{10}$.** Since we need to find the probability that a student has *at least 1 pencil*, this will include all of the students other than the students who had 0 pencils. Since there were 40 students in total, and 12 of them had 0 pencils in their bags, this means that 28 of them had 1 or more pencils in their bags. Therefore, the probability that a student chosen at random will have at least 1 pencil is $\frac{28}{40} = \frac{7}{10}$.

462. **The correct answer is choice c.** The formula for the surface area of a sphere is $SA = 4\pi r^2$. If the kickball has a radius of 4.5 inches, then the equation becomes $SA = 4\pi(4.5 \text{ in.})^2 = 254.34 \text{ in.}^2$.

Choice **a** is incorrect because it is a rounded version of choice **b**.

Choice **b** is incorrect because it results from only finding the area of a circle, $\pi r^2 = 63.585 \text{ in.}^2$.

Choice **d** is incorrect because it is rounded too high to be an approximate surface area. Choice **c** is a much more accurate choice.

463. The correct answer is $\frac{1}{41}$. The probability of picking a blue candy the first time is $\frac{7}{42}$. What is the probability of picking a blue candy the next time? After one piece is taken out, there are six blue candies left and 41 candies left overall, which is $\frac{6}{41}$. The compound probability of picking two blue candies in a row is found by multiplying the probability of the two events together:

$$\frac{7}{42} \times \frac{6}{41}$$
$$\frac{7}{7 \times 6} \times \frac{6}{41}$$
$$\frac{\cancel{7}}{\cancel{7} \times \cancel{6}} \times \frac{\cancel{6}}{41}$$
$$\frac{1}{1 \times 1} \times \frac{1}{41}$$
$$\frac{1}{41}$$

464. The correct answer is choice b. Mitch has six songs from which to choose three songs. It is important to understand that the order of the songs does not matter. The group of songs 1, 2, and 3 is the same as the group 1, 3, 2; 2, 3, 1; 2, 1, 3; 3, 1, 2; and 3, 2, 1. These are all the same group of three songs—they do not count as six different groupings.

Use the combination formula to solve the problem:

$$_nC_r = \frac{n!}{r!(n-r)!}$$

Mitch has six songs and he chooses three:

$$_6C_3 = \frac{6!}{3!(6-3)!}$$
$$_6C_3 = \frac{6!}{3!3!}$$
$$_6C_3 = \frac{6 \times 5 \times 4 \times \cancel{3} \times \cancel{2} \times \cancel{1}}{3 \times 2 \times 1 \cdot \cancel{3} \times \cancel{2} \times \cancel{1}}$$
$$_6C_3 = 20$$

465. The correct answer is choice c. The mode is the number that occurs most often in a set of data. In these data, the number 7 appears three times, more than any other number.

Choice a is incorrect because 360 appears only once in the data set and is therefore not the mode.

Choice b is incorrect because 100 appears only twice and is therefore not the mode.

Choice d is incorrect because 1 appears only once in the data set and is therefore not the mode.

466. **The correct answer is choice c.** The probability of an event (picking a red jelly bean) is found using the following formula:

$$P(\text{event}) = \frac{\text{favorable}}{\text{possible}}$$

In this problem, the number of favorable events is 7 because there are 7 red jelly beans. The number of possible events is 21 because there are 21 jelly beans in the bag and any one of those jelly beans can be chosen.

$$P(\text{red}) = \frac{7}{21} = \frac{1}{3}$$

The probability of drawing a red jelly bean is $\frac{7}{21}$, which can be reduced to $\frac{1}{3}$.

Choice **a** is incorrect because you incorrectly added the total number of jelly beans as 20, not 21.

Choice **b** is incorrect because the numerator should be the total number of red jelly beans, or 7.

Choice **d** is incorrect because there are only 7 red jelly beans out of a total of 21: this is less than $\frac{1}{2}$ of the total number of jelly beans.

467. **The correct answer is choice d.** $430 + 425 + 345 = 1,200$ parts are accounted for. Since the total is 1,650: $1,650 - 1,200 = 450$ blue parts. When randomly picking a part, the chance of getting blue part is 450 out of $1,650 = \frac{450}{1,650}$. Simplify the expression:

$$\frac{450}{1,650} \div \frac{150}{150} = \frac{3}{11}.$$

468. **The correct answer is $\frac{1}{1,000}$.** The probability of an event happening is $P(E) = \frac{\text{\# of desirable events}}{\text{Total \# of possible events}}$. Since there are 1,000 possible three-digit ID numbers from 000 to 999, and the only ID number that Delilah wants is 314, her chances of getting her π ID number are $\frac{1}{1,000}$.

469. **The correct answer is $\frac{1}{100}$.** There are 10 possible ID numbers that are triple repeating digits: 000, 111, 222, . . . 999. There are 1,000 possible events of three-digit numbers from 000 to 999. Since the probability of an event happening is $P(E) = \frac{\text{\# of desirable events}}{\text{Total \# of possible events}}$, the probability that Skye will get a triple repeating digit ID number is $\frac{10}{1,000} = \frac{1}{100}$.

470. **The correct answer is $\frac{2}{125}$.** First, we will list out all of the ID numbers that have consecutive digits counting up: 012, 123, 234, 345, 456, 567, 678, and 789. Since there are 8 that count upward, that means there must be 8 that count downward, so there are 16 ID numbers in total that Lulu would be happy to have. Since there are 1,000 possible three-digit numbers from 000 to 999, the probability that she'll get one of these 16 numbers is $\frac{16}{1,000} = \frac{2}{125}$.

471. **The correct answer is choice c.** Use the counting principle to solve this problem. Since there are three ways to choose the main item, three ways to choose a candy, and five ways to choose a drink, there are a total of $3 \times 3 \times 5 = 45$ different ways to choose a combo.

472. **The correct answer is choice c.** The order of the five students does not matter in this situation because there are not awards for first, second, third, fourth, or fifth place. Therefore, use the formula for finding the number of combinations when order does not matter: $\frac{n!}{r!(n-r)!}$, where n = the number of students and r = the number of students chosen (i.e., $n = 23$ and $r = 5$). Substitute into the equation and solve for the number of combinations that could result from five of 23 students getting recognized each week:
$$\frac{n!}{r!(n-r)!} = \frac{23!}{5!(23-5)!} = \frac{23 \times 22 \times 21 \times 20 \times 19}{5 \times 4 \times 3 \times 2 \times 1} = \frac{4{,}037{,}880}{120} = 33{,}649$$
Choice **a** is incorrect because this is just the number of students chosen each week.

Choice **b** in incorrect because this is 5!, which is only a part of the equation.

Choice **d** is incorrect because this is the number of combinations if this were a permutation problem (i.e., if the order mattered).

473. **The correct answer is choice d.** There is a total of 150 attendees. Those whose age is greater than 40 are those in the bottom two rows, which gives 52 out of 150 attendees. So the probability of randomly choosing an attendee whose age is greater than 40 is $\frac{52}{150} = \frac{26}{75}$.

Choice **a** is incorrect because you divided the correct number (52) by 100 instead of the total number of attendees, 150.

Choice **b** is incorrect because this uses only the last row; you should use the bottom *two* rows.

Choice **c** is incorrect because this is the probability that a randomly selected attendee's age is *not* greater than 40.

474. **The correct answer is choice a.** The median of this data set is the average of the 12th and 13th outcomes, namely $\frac{6+8}{2} = 7$.

Choice **b** is incorrect because this middle number listed in the left column is not automatically the median; you must account for the different frequencies.

Choice **c** is incorrect because this is the average of the 13th and 14th outcomes, but since there are 24 outcomes, you should have averaged the 12th and 13th outcomes.

Choice **d** is incorrect because this is the mode, not the median.

475. **The correct answer is 92.33%.** To find Katie's final grade, a weighted average must be calculated since Mr. Gallespie weights each type of assignment differently. Find the average score of each type of assignment, multiply it by its percentage weight, then add the percentages together to get Katie's final grade.

$$\text{Final grade} = 0.60\left(\tfrac{93 + 97 + 88 + 91 + 95}{5}\right) + 0.15\left(\tfrac{90 + 99 + 100 + 95 + 96}{5}\right)$$
$$+ 0.25\left(\tfrac{89 + 90 + 85 + 92}{4}\right)$$
$$= 0.60(92.8) + 0.15(96) + 0.25(89)$$
$$= 55.68 + 14.4 + 22.25$$
$$= 92.33$$

476. **The correct answer is choice b.** Let x represent the number of points he must score in the seventh game. Compute the average as follows:

$$\frac{48 + 38 + 38 + 34 + 44 + 44 + x}{7} = 40$$
$$\frac{246 + x}{7} = 40$$
$$x = 280 - 246 = 34$$

So he must score 34 points.

Choice **a** is incorrect. This is the average of the first six games; scoring this many points in the seventh game would yield an average of 41, not 40.

Choice **c** is incorrect because this would ensure that the mode is 44, but the mean would be higher than 40.

Choice **d** is incorrect because this would ensure that the mode is 38, but the mean would be slightly higher than 40.

477. **The correct answer is choice d.** Spins are independent of each other. So no matter how many times you were to spin the spinner, the probability would always be the same.

Choice **a** is incorrect because this is the probability of getting red.

Choice **b** is incorrect because there are not equal numbers of red and green spaces, so red and green are not equally likely.

Choice **c** is incorrect because this is the probability of getting two greens in two spins, which is not the event for which the probability was asked.

478. **The correct answer is choice d.** Note that the color and the type of bike seat do not factor into how the bike is chosen when asking for the probability that the bike chosen is a mountain bike. So, since there is an equal likelihood of choosing any of the three types of bikes, the probability of randomly choosing a mountain bike is $\frac{1}{3}$.

Choice **a** is incorrect because this would insist on a specific structure, color, and type of bike seat.

Choice **b** is incorrect because this would insist on a specific type of seat.

Choice **c** is incorrect because this would insist on a specific color.

479. **The correct answer is choice b.** Put the nine ring scores in order from least to greatest. The middle value (the fifth value) is the median score:

3, 3, 3, 4, **4**, 5, 6, 8, 9

The median score is 4.

Choice **a** is incorrect because you confused the mode with the median. The mode is the most common data entry, but the median is the middle data entry when the data is listed from least to greatest.

Choice **c** is incorrect because you confused the game numbers for the data and you selected the middle game number, when you were asked to find the median *score*.

Choice **d** is incorrect because although it is the middle number of ring scores in the table as they are listed, the data must first be arranged from least to greatest before determining the median, and you forgot this critical step.

480. **The correct answer is choice d.** Arrange the data in increasing order. The median of the original data set is 168, and replacing 320 by 350 does not change this value since 168 remains the middle value in the data set. The mean, however, will increase since the sum of the data has increased.

Choices **a** and **c** are incorrect because the median remains unchanged since it is concerned with position within the data set, and replacing 320 by 350 does not change this.

Choice **b** is incorrect because the mean will increase since it is obtained by summing the values and dividing by the same number, 7.

481. **The correct answer is $\frac{1}{13}$.** Since the first utensil removed is not returned to the drawer, the number of utensils in the drawer decreases to 12. When calculating compound probability, it is always assumed that the first event is a success, in order to determine the simple probability of the second event. This means that P(first utensil fork) = $\frac{4}{13}$ and P(second utensil fork) = $\frac{3}{12}$. Multiply both of these simple probabilities together to find the compound probability of both events happening: $\frac{4}{13} \times \frac{3}{12} = \frac{12}{156} = \frac{1}{13}$.

482. **The correct answer is 9%.** Since these students are being picked without replacement, each time a student is picked, the class size will decrease by 1. When calculating "without replacement" probability, you must assume that each event is a success so the number of boys available to pick from must also reduce by 1:

P(first student being a boy) = $\frac{14}{30}$

P(second student being a boy) = $\frac{13}{29}$

P(third student being a boy) = $\frac{12}{28}$

For compound probability, multiply all three of these probabilities together:

P(boy then boy then boy) $\frac{14}{30} \times \frac{13}{29} \times \frac{12}{28} = \frac{2,184}{24,360} = 0.09 = 9\%$

So, the probability that all three students chosen at random will be boys is 9%.

483. **The correct answer is choice a.** Of all of the scatter plots with best fit lines shown, this particular one correctly illustrates the rising from left to right pattern of the data points and minimizes the vertical distance between the points and the line itself.

Choices **b** and **d** are incorrect because these lines do not accurately illustrate the rising from left to right trend evident in the data points.

Choice **c** is incorrect because while this correctly illustrates the rising from left to right trend in the data points, it lies beneath all the points and therefore does not minimize the vertical distance between the points and the line.

484. **The correct answer is choice c.** Let x be the score in the fifth game. We need the range of x for which the following inequality is satisfied:

$$200 \le \frac{191 + 218 + 210 + 171 + x}{5} \le 215$$

Solving this inequality yields

$$200 \le \frac{790 + x}{5} \le 215$$
$$1{,}000 \le 790 + x \le 1{,}075$$
$$210 \le x \le 285$$

So, the necessary range is 210 to 285.

Choice **a** is incorrect because this cannot work since the average of the first four games is 197.5.

Choice **b** is incorrect because you divided by 4, not 5.

Choice **d** is incorrect because this is the range that describes her first four games.

485. **The correct answer is choice a.** There are 5 vowels and 9 consonants in the bag, so it is not equally likely to choose a vowel or a consonant. The statement is false.

Choice **b** is an incorrect answer because 7 of the 14 tiles (50%) in the bag satisfy this condition, so this statement is true.

Choice **c** is an incorrect answer because there are 2 P tiles and 2 B tiles, so this statement is true.

Choice **d** is an incorrect answer because there are 4 E tiles and 2 C tiles, so the statement is true.

486. **The correct answer is choice b.** The spinner has ten equal sections, so the probability of the spinner landing on any one number is $\frac{1}{10}$. There are four numbers on the spinner that are less than 5 (1, 2, 3, and 4). The probability of the spinner landing on a number less than 5 is $\frac{4}{10}$, or $\frac{2}{5}$.

Choice **a** is incorrect because a $\frac{1}{5}$ probability is the same as $\frac{2}{10}$, but there are 4 numbers less than 5 on the spinner, not just 2.

Choice **c** is incorrect because you included 5 in the number of successful events; however, the question asked about the spinner landing on a number *less than* 5. Since there are only 4 numbers *less than* 5 on the spinner the probability is $\frac{4}{10}$ or $\frac{2}{5}$.

Choice **d** is incorrect because you calculated the probability of the spinner landing on the *number 5* instead of on a number *less than 5*. Remember that there are 4 numbers less than 5 on the spinner, so the probability must be $\frac{4}{10}$ or $\frac{2}{5}$.

487. **The correct answer is choice a.** Before Evander removes a coin, there are 16 coins in the bank:

3 quarters + 5 pennies + 2 nickels + 6 dimes = 16 coins

After Evander removes the first quarter, there are:

2 quarters + 5 pennies + 2 nickels + 6 dimes = 15 coins

There are only 15 coins in the bank now, and only two of them are quarters. If Evander selects another coin, the probability that it will be a quarter is $\frac{2}{15}$.

Choice **b** is incorrect because you reduced the total number of coins to 15, but you forgot to reduce the number of quarters from 3 to 2.

Choice **c** is incorrect because you reduced the number of quarters to 2 but you forgot to reduce the number of coins from 16 to 15.

Choice **d** is incorrect because you selected the probability that the first coin would be a quarter. Since the first quarter was not returned to the bank, there were only 2 quarters available out of 15 total coins, for the second coin pull.

488. **The correct answer is choice b.** The median of seven numbers is the middle number after the numbers are listed in order from least to greatest (or from greatest to least). So, the median of this set of numbers is 11, which is the fourth number in the ordered list:

–3, 7, 8, 11, 16, 22, 30

Choice **a** is incorrect. It is the least value in the data set and therefore is not a middle number.

Choice **c** is incorrect because it is the mean of the data, not the median.

Choice **d** is incorrect because it is not the middle value and therefore cannot be the median.

489. **The correct answer is 286.** Since order does not matter for this problem, use the combinations formula to see how many different groupings of appetizers Mahershala and his friends can order. Set $n = 13$ and $k = 3$:

$$_nC_r = \frac{n!}{r!(n-r)!}$$
$$_{13}C_3 = \frac{13!}{3!(13-3)!}$$
$$_{13}C_3 = \frac{13!}{3!(10)!}$$
$$_{13}C_3 = \frac{13 \times 12 \times 11 \times \cancel{10} \times \cancel{9} \times \cancel{8} \times \cancel{7} \times \cancel{6} \times \cancel{5} \times \cancel{4} \times \cancel{3} \times \cancel{2} \times \cancel{1}}{3 \times 2 \times 1 \cdot (\cancel{10} \times \cancel{9} \times \cancel{8} \times \cancel{7} \times \cancel{6} \times \cancel{5} \times \cancel{4} \times \cancel{3} \times \cancel{2} \times \cancel{1})}$$
$$_{13}C_3 = 286$$

490. **The correct answer is $\frac{1}{12}$.** Since there are two possible outcomes with the coin toss and six possible outcomes with rolling a die, multiply these two numbers together to calculate the total number of possible combinations: $2 \times 6 = 12$. There is only one way that Nadezda can both get a tails when she flips the coin *and* get a 3 when she rolls the die. Since there are 12 possible outcomes in total, the probability that she will arrive at this one specific outcome is $\frac{1}{12}$. Another way to approach this is to multiply both of the individual probabilities together. The probability of getting tails is $\frac{1}{2}$ and the probability of rolling a 3 is $\frac{1}{6}$, so the combined probability of both of these events happening is their product: $\frac{1}{2} \times \frac{1}{6} = \frac{1}{12}$.

491. **The correct answer is choice a.** Since there are different numbers of students in the categories, it is necessary to find the weighted average. Begin by calculating the total number of miles traveled one way by *all* of Ms. Bradley's students by multiplying the number of miles in the first column by the number of students in the second column. Write these products in a third column:

NUMBER OF MILES TRAVELED ONE WAY	NUMBER OF STUDENTS	(# MILES)(# STUDENTS) = SUBTOTALS OF MILES
1	4	1(4) = 4
2	3	2(3) = 6
3	5	3(5) = 15
4	6	4(6) = 24
5	3	5(3) = 15

Total number of one-way miles: $4 + 6 + 15 + 24 + 15 = 64$. Then divide that by the number of total students in her viola class: $4 + 3 + 5 + 6 + 3 = 21$ students. So, the average miles driven will be total miles divided by number of students: $\frac{64}{21} \approx 3.05$. On average, Ms. Bradley's students travel about 3.1 miles each way to get to her viola class.

Choice **b** is incorrect because you had a calculator error while finding the subtotals of miles reported by the students.

Choice **c** is incorrect because 4 miles is not the average.

Choice **d** is incorrect because you added the number of student responses together and divided that by the 5 different answer choices, but this is not how to calculate a weighted average.

492. **The correct answer is $7.96/can.** Since the different cans of fruit have different prices, each type must get multiplied by the quantity that was purchased before the average can be found. Since Stephanie bought 20 cans at $6.50 each and 28 cans at $9 each, the total amount spent can be set up as $6.50(20) + $9.00(28) and the total number of cans will be found by adding 20 + 28:

$$\text{Mean price per can} = \frac{\text{Total cost of all cans of fruit}}{\text{\# of cans of fruit}}$$

$$\text{Mean price per can} = \frac{\$6.50(20) + \$9.00(28)}{20 + 28}$$

$$\text{Mean price per can} = \frac{\$382}{48} = \$7.96$$

So, the average price per can of fruit that Stephanie spent was $7.96/can.

493. **The correct answer is choice d.** Finding the odds of an event happening is different from finding the probability of something happening. While probability compares the number of successful events to the number of total events, the odds of an event happening are the ratio of the number of successful events to the number of unsuccessful events. Since there are 27 students, and 6 of them are in group A, that means that 21 of them are *not* in group A. The odds of drawing a student from group A are number of ways to get group A versus number of ways to *not* get group A. This is 6 to 21, which reduces to 2 to 7.

Choice **a** is incorrect because you found the odds of Mr. Syed choosing a student who was *not* in group A. The first number must represent the number of successful events.

Choice **b** is incorrect because you found the *probability* of Mr. Syed choosing a student from group A, and not the *odds*.

Choice **c** is incorrect because you found the *probability* of Mr. Syed choosing a student who was not in group A, and not the *odds* of selecting a student who *was* in group A.

494. **The correct answer is 60 combinations.** The fundamental counting principle states that the total number of possible outcomes for the three separate events is the product of the numbers of possibilities for each event. Since Maya has four favorite salads, five favorite entrées, and three favorite desserts, multiply 4 by 5 by 3 to get 60 unique combinations.

495. **The correct answer is choice c.** A probability is written as a fraction. The numerator is the number of outcomes that make an event true, and the denominator is the number of possible outcomes. The event that Isabel selects a card that is *not* blue is true when Isabel selects a yellow, green, purple, or red card. Find the total number of yellow, green, purple, and red cards:

> 10 yellow cards + 10 green cards + 3 purple cards + 2 red cards = 25 cards

The number of possible outcomes when Isabel selects one card from the deck is equal to the number of cards in the deck:

> 10 yellow cards + 10 green cards + 5 blue cards + 3 purple cards + 2 red cards = 30 cards

Finally, write the probability as a fraction:

$$\frac{\text{\# of outcomes that make the event true}}{\text{Total \# of outcomes}} = \frac{25}{30} = \frac{5}{6}$$

Choice **a** is incorrect because it is the probability that the card *is* blue. Choice **b** is incorrect because you likely added the total number of non-blue cards, subtracted the number of blue cards (5), and put it over 25. You must add the TOTAL number of cards and put the total number of non-blue cards over this value. Choice **d** is incorrect because you subtracted the 1 card selected and put this value over the total number of cards.

496. **The correct answer is choice c.** This is a compound probability problem. The probability of the first event happening—grabbing a square—is $\frac{6}{22}$. The probability of the second event happening—grabbing a triangle after a square is already removed—is $\frac{9}{21}$. Thus, the compound probability of both events happening is found by multiplying: $\frac{6}{22} \times \frac{9}{21} = \frac{54}{462} = \frac{9}{77}$.

Choice **a** is incorrect because it is the probability of the first event only.

Choice **b** is incorrect because it is the probability of the second event only.

Choice **d** is incorrect because it is not in lowest terms.

497. **The correct answer is choice b.** Since only members of the year-book staff were asked the question, and such a group might naturally tend to like more writing-focused courses, the sample is not representative of the entire junior class.

Choice **a** is incorrect because the subjects themselves are not as relevant as being asked to select one of them.

Choice **c** is incorrect because he is trying to draw a conclusion about juniors at his own school.

Choice **d** is incorrect because you do want to get as inclusive a sample as possible, which would include those on the yearbook staff. The point, though, is that you do not want to include just people on the yearbook staff in the sample.

498. **The correct answer is choice d.** Use the addition formula because the events share common outcomes:

$$P(\text{male or spotted}) = P(\text{male}) + P(\text{spotted}) - P(\text{male and spotted})$$
$$= \frac{6}{9} + \frac{4}{9} - \frac{3}{9}$$
$$= \frac{7}{9}$$

Choice **a** is incorrect because this is the probability that the puppy selected is a spotted female.

Choice **b** is incorrect because this is the probability that the puppy selected is a spotted male.

Choice **c** is incorrect because this is the probability that the puppy selected is spotted.

499. **The correct answer is choice c.** Using the fundamental counting principle, simply multiply the three numbers together:
$6 \times 9 \times 11 = 594$.

Choice **a** is incorrect because it simply lists the number of total items but does not consider the combinations of the items.

Choice **b** is the result of improper calculation.

Choice **d** is incorrect because it adds in another group of wrapping paper/ribbon combinations: $6 \times 9 \times 11 + (9 \times 11) = 693$.

500. **The correct answer is choice b.** If you divide the circular spinner into sixths, you can see that the number 1 occupies four-sixths of the circle. Since $\frac{4}{6} = \frac{2}{3}$, the probability of spinning the number 1 is $\frac{2}{3}$.

Choice **a** is incorrect because there are 5 parts of the spinner, and 3 of those parts have a 1 on them.

Choice **c** is incorrect because you likely used the total number of 1s as your denominator, but the denominator must represent all parts of the spinner.

Choice **d** is incorrect because there are 5 total parts of the spinner, not 4.

501. **The correct answer is choice b.** The number of hearts in one deck is 13, so there are 26 in the combined deck. The number of aces in one deck is 4, so there are 8 in the combined deck. There are 2 aces of hearts in the combined deck, and we do not want to count them twice. Hence, the desired probability is $\frac{26 + 8 - 2}{104} = \frac{32}{104} = \frac{4}{13}$.

Choice **a** is incorrect because this is the probability of selecting neither a heart nor an ace.

Choice **c** is incorrect because you counted the 2 aces of hearts twice.

Choice **d** is incorrect because this is the probability of getting a heart, but you did not account for the possibility of getting an ace.

Appendix: Formula Reference Sheet

The following are the formulas you will be supplied with on the GED® Mathematical Reasoning test.

Area

Parallelogram: $A = bh$

Trapezoid: $A = \frac{1}{2}h(b_1 + b_2)$

Surface Area and Volume

Rectangular/right prism:	$SA = ph + 2B$	$V = Bh$
Cylinder:	$SA = 2\pi rh + 2\pi r^2$	$V = \pi r^2 h$
Pyramid:	$SA = \frac{1}{2}ps + B$	$V = \frac{1}{3}Bh$
Cone:	$SA = \pi rs + \pi r^2$	$V = \frac{1}{3}\pi r^2 h$
Sphere:	$SA = 4\pi r^2$	$V = \frac{4}{3}\pi r^3$

(p = perimeter of base B; $\pi \approx 3.14$)

Algebra

Slope of a line: $m = \frac{y_2 - y_1}{x_2 - x_1}$

Slope-intercept form of the equation of a line: $y = mx + b$

Point-slope form of the equation of a line: $y - y_1 = m(x - x_1)$

Standard form of a quadratic equation: $y = ax^2 + bx + c$

Quadratic formula: $x = \frac{-b \pm \sqrt{b^2 - 4ac}}{2a}$

Pythagorean theorem: $a^2 + b^2 = c^2$

Simple interest: $I = prt$

(I = interest, p = principal, r = rate, t = time)

Additional Online Practice

Using the codes below, you'll be able to log in and access additional online practice materials!

Your free online practice access codes are:
FVEAB2TP5RD5D16QDW51
FVE5228375443I7PX6IV

Follow these simple steps to redeem your codes:
- Go to **www.learningexpresshub.com/affiliate** and have your access codes handy.

If you're a new user:
- Click the New user? Register here button and complete the registration form to create your account and access your products.
- Be sure to enter your unique access code only once. If you have multiple access codes, you can enter them all—just use a comma to separate each code.
- The next time you visit, simply click the **Returning user? Sign in** button and enter your username and password.
- Do not re-enter previously redeemed access codes. Any products you previously accessed are saved in the **My Account** section on the site. Entering a previously redeemed access code will result in an error message.

If you're a returning user:
- Click the **Returning user? Sign in** button, enter your username and password, and click **Sign In**.
- You will automatically be brought to the **My Account** page to access your products.
- Do not re-enter previously redeemed access codes. Any products you previously accessed are saved in the My Account section on the site. Entering a previously redeemed access code will result in an error message.

If you're a returning user with a new access code:
- Click the **Returning user? Sign in** button, enter your username, password, and new access code, and click **Sign In**.
- If you have multiple access codes, you can enter them all—just use a comma to separate each code.
- Do not re-enter previously redeemed access codes. Any products you previously accessed are saved in the **My Account** section on the site. Entering a previously redeemed access code will result in an error message.

If you have any questions, please contact Customer Support at Support@ebsco.com. All inquiries will be responded to within a 24-hour period during our normal business hours: 9:00 a.m.–5:00 p.m. Eastern Time. Thank you!

Notes

Notes